7-66

*Ruth Tuthill Hawes
Endowed Humanities
Fund for History*

THE LIBRARY FOUNDATION

Serving the People of Multnomah County

THE PATHBREAKERS
from RIVER *to* OCEAN

THE STORY OF THE GREAT WEST FROM THE TIME OF CORONADO TO THE PRESENT

GRACE RAYMOND HEBARD, Ph. D.

PROFESSOR OF POLITICAL ECONOMY, STATE UNIVERSITY OF WYOMING,
AUTHOR OF "HISTORY AND GOVERNMENT OF WYOMING."

Four Maps and Numerous Illustrations

THIRD EDITION

CHICAGO
THE LAKESIDE PRESS
1913

The Lakeside Press
R. R. DONNELLEY & SONS COMPANY
CHICAGO

B. L. Zimm, Sculptor

THE WHITE MAN'S PILOT, SACAJAWEA, AND BAPTISTE, WHO ACCOM-
PANIED LEWIS AND CLARK ACROSS THE CONTINENT

TO MY FRIEND,
DOCTOR AGNES M. WERGELAND
A PATHBREAKER

PREFACE

Multitudes of books have been written for pupils of our schools recording the valiant deeds of the explorers who have made their field of operation east of the Mississippi. De Soto, Smith, Marquette, Clark, Boone and the many adventurous heroes who plied up and down all of the streams between the mighty river and the ocean to the East, have received, each in turn, due attention, and their deeds have not only been recorded upon the pages of books but written in the hearts of the American youths. The West, or that land situated between the Mississippi and the western coast, has not received its due attention in school book form. To enable the future citizens, particularly those who live in the states carved out of this story-making territory, to familiarize themselves with the brave deeds of these earliest inhabitants in an unsettled and unorganized territory is the purpose of this publication. No territory or period of history so abounds in heroic deeds, daring adventures, and hazardous enterprises which have directly served to bring about civilization as the region known as the Great West. The tale is not only interesting but fascinating from the earliest beginning to the present day. The turbulent streams, the rugged and forbidding mountains, the limitless plains, the hostile natives, and the extremes of climate made the struggle a hard one, and demanded men of courage who had faith in themselves and the object of their conquests. The wonderful story is too long to appear between the covers of any one book, yet the hope is expressed that the facts assembled may awaken a new interest in the labors of those untiring climbers of streams and mountains who made that undeveloped country a part of our present-day possessions. If this is accomplished the labor of preparing the book will be abundantly rewarded.

The author is indebted for valuable corrections and suggestions to

Mrs. Eva Emery Dye, Messrs. O. D. Wheeler, Owen Wister, and Emerson Hough, whose literary works have done so much to create a just estimate of the West and its frontier life; but on these authors rests no responsibility for any inaccuracies that may be found in the text. Further mention of gratitude should be made to Mr. E. F. McGinnis for helpful ideas in outlining the work; to Mr. Frank Bond, Chief of the Drafting Division of the General Land Office, for assistance in the preparation of the trail maps; and to co-workers, Dr. A. M. Wergeland and Dr. J. E. Downey, for encouragement in the work and for friendly criticism.   GRACE RAYMOND HEBARD.

UNIVERSITY OF WYOMING,
   LARAMIE, July 2, 1911.

CONTENTS

THE PATHBREAKERS
from RIVER *to* OCEAN

THE PATHBREAKERS FROM RIVER TO OCEAN

CHAPTER I

THE EARLY EXPLORERS

1. CORONADO
2. THE VERENDRYES
3. LEWIS AND CLARK
4. ZEBULON PIKE

1. CORONADO

No narrative of the early explorers of southwestern North America would be complete without some mention of Coronado, Francisco Vasquez de Coronado, seeker of the "Seven Cities of Cibola." To prepare the way for an account of his exploits it is needful that you recall the journeys of Cordova to Central America and the quantities of golden ornaments he found among the natives there, the wanderings of Narvaez from Florida northwestward, the escape of his treasurer, Cabeza de Vaca, his wanderings through what is now Texas, his return to the Spaniards in Mexico with tales of cities of marvelous wealth in the interior, and, finally, the conquest of Mexico by the brilliant Cortez.

In 1539, Coronado was made provisional governor of Nueva Galicia (New Gaul), by Antonio de Mendoza, viceroy of Mexico. This Viceroy Mendoza, who had been filled with enthusiasm over the accounts that Cabeza de Vaca had brought home, urged Coronado to take charge of his province at once and to explore the unknown country to the north immediately. Coronado, with the spirit of ad-

1

venture in his blood, was equal to the task and was eager to be off.

The army was financed from the personal wealth of Coronado and what he could borrow, though the command of the soldiers was granted to him by the viceroy. The expense of equipping the expedition was about two hundred and fifty thousand dollars, or sixty thousand ducats. If Coro-

From an old cut
THE STYLE OF CANNON USED BY CORTEZ AND CORONADO

nado did not succeed in his high ambition,— the loss was his. If he won,— he was dizzy with the vision of the empires he was to conquer and own. He would have wealth greater than that of Cortez, and lands unlimited.

Early in 1540, piloted by a monk, Fray Marcos, who had previously penetrated to the Zuni Villages, he started with three hundred horsemen, foot soldiers, crossbow men, arquebusiers, eight hundred Indians, and a thousand extra horses for ammunition and baggage.

Northward through tribes more or less hostile they marched, near the present site of Tombstone, Arizona, on to

Salt River, north across the Mogollon Mountains, then north-east to the Little Colorado. Here the first station of their travels was reached, the Zuni Pueblos. We are told that the city of Zuni is the home of a people who lived there centuries before the coming of Columbus. "There they still live, with very little change. The march of progress that has swept away other Indian tribes has spared the

ANCIENT CLIFF RUINS NEAR SANTA FÉ

lonely little pueblo communities in their adobe terraced houses, surrounded by the arid deserts." [1]

These adobe terraces made excellent forts. Rising tier on tier to the height of three or sometimes four stories, with no doors, they covered with unbroken walls many acres, and presented a formidable front to the little army destitute of cannon with which to batter a breach. Ingress to these houses is had through the roof by aid of ladders. Upon the approach of the intruders, the Zuni had drawn up all their

[1] Johnson. Pioneer Spaniards of North America. Little, Brown & Co.

ladders and ranged themselves on the terraces intent on defending their homes.   Wood for the construction of new ladders was hard to obtain, and when the means of assault were finally provided it was no child's play to storm that fortress through the hail of arrows and stones from the warriors in the terraces.   Coronado's shining armor and his foremost

George Wharton James, "Indians of the Painted Desert Region"
PUEBLO HOMES

place in the assault made him an especial target.   After the place was won he had gaping wounds on his face, an arrow in one of his feet, and many stone bruises on his legs and arms, and tells us that if it had not been for the strength of his armor "it would have gone hard with me."   But more bitter than the perils of the assault was the disappointment of the victors upon finding that here was no gold or precious stones,— none of the wealth they had marched and starved, fought and bled, to win.   Evidently these were not the famed "Seven Cities of Cibola."   They must be still farther on.

So the conquerors passed on; scouts were sent in various

directions, all bringing back similar reports of "no cities
of gold." In August of this year, 1540, the famous Grand
Cañon of the Colorado was discovered by a division of
the expedition under the command of Garcia Lopez de
Cardenas. One often wonders if this Cañon were the real
"city of gold," for the discovery and possession of which

From Thevet's Les Singularitez
THE EARLIEST KNOWN PICTURE OF A BUFFALO

more than one continent wasted its blood and treasure. It
would take no great stretch of the imagination to fancy that
the glittering and sparkling mica-bearing formations re-
sponding to the sun's rays were houses built of gold. The
tradition of this city was not a dream. Some one had observed something. For want of better description the vision
was called a "city of gold." Was this the end of the rainbow
with its proverbial "pot of gold"?

From this point the party pushed eastward. Had they
gone west might they not have outstripped by three hundred
years the "forty-niners" in their mad rush for gold? An
Indian guide, named "Turk" by the Spaniards, told mar-

velous tales of the untold wealth of the city of Quivira to the
northeast. It was towards this city they were now pushing.
We next find the army in the Staked Plains of Texas, or the
Buffalo Plains. On these wide plains they encountered the
"humpbacked oxen." These buffalo had first been accurately described to the European world by Vaca. They were

RUINS OF ANCIENT CLIFF DWELLERS, NEAR SANTA FÉ

also called "cattle or cows of Cibola." Coronado obtained
many buffalo robes from the inhabitants of his first conquered village, and carried them to Spain on his return
voyage. From here the band of explorers pushed up north
through the land occupied now by Oklahoma, into Kansas,
well toward the north central part of the state. After traveling for forty-five days with thirty of his best mounted
horsemen, Coronado decided that it was useless to go farther.
He had found the city a village of straw huts, the people

more savage than any other they had met. The sight of these villages and of the immense plains with their countless herds of "horned oxen" was the only reward of his labors. There was no alternative but to return. Defeated and dejected the little army journeyed back to Mexico, presenting a picture not unlike that of the defeated Napoleon and his army on their return from Russia. Less than half of the invaders returned to tell the story of defeated ambition, one hundred and fifty leaving their bones to bleach on the plains and the mountain sides.

In the fall of 1542 the explorer arrived at the City of Mexico "very sad and very weary, completely worn out and shamefaced." Was the expedition a failure? He had not found a single thing for which he had sought. He returned empty handed; fame and money were gone. But History will say that he did much toward the finding of that vast country east of the Colorado River from the Grand Cañon to Gulf of California, and all of those countless miles of prairie extending northward to the southern boundary of Nebraska, and possibly even into that State. To Coronado must we accord the place of the first "early explorer" of those lands situated between the Mississippi River and the Pacific Ocean.

2. THE VERENDRYES

Two centuries after Coronado had made his unsuccessful attempt a French Canadian and his sons endeavored to find a northwest passage to the Mer' de l'Ouest, or Pacific Ocean. The journey was not to be one entirely of discovery and conquest for the French nation; combined with the desire for exploration there was the purpose to grow wealthy through the finding of new and rich trapping-grounds.

At the beginning of the eighteenth century there were persistent rumors of a river that flowed toward the Sea of the West. The Indians pointed toward the setting sun, declaring that there was a river that throbbed with a pulse that was frightening just before it emptied into a Great Lake, the waters from which were salty and undrinkable. They told further of a strange people who wore iron dresses and rode on horseback. But there was no exact knowledge of this mighty westward-flowing stream, nor of those strange iron-dressed warriors. The river was doubtless the Columbia, known by the natives before the time of Lewis and Clark. The warriors? Might they not have been some of the Spaniards from Mexico who, more daring than the others, had pushed up farther north than any page of history records? At this time the Western Sea was supposed to lie somewhere between America and Japan. The Pacific Ocean was supposed to limit the western boundary of the domain of the Sioux Indians, but no definite information could be obtained. Everyone was questioned on the subject but with no satisfactory results. "The missionaries and the officers had nothing to tell; the voyagers and Indians knew no more than they, and invented confused and contradictory falsehoods to hide their ignorance." [1]

By the authority of the Duke of Orleans, companies were allowed to organize to attempt to discover the long-sought route to the Pacific. Their duties were twofold: to establish missions among the Sioux where the missionaries could learn the language of the Indians and thus get more information about the unknown, hazy sea; and to establish a monopoly of the Sioux fur trade. After the failure of others it was left to Pierre Gaultier de Varrennes de la Verendrye and his sons to explore to the west as far as the Big Horn Mountains. Pierre was the son of René Gaultier de Varrennes, a trader in furs near the head of Lake Superior. He took in addition the name La Verendrye. The center of his trading was north of Lake Superior, in which locality he heard the tradition of the *Western Sea*. "These people are liars," he is

[1] Parkman. A Half-Century of Conflict. Little, Brown & Co.

reported to have said, "but now and then they tell the truth." He petitioned the King of France, Louis XV, for aid in organizing a company to search for this ocean. This aid could not be obtained, but Verendrye was authorized to build forts along the route, and to have a monopoly of the fur trade with the Indians who inhabited the territory toward the Western Sea. Knowing to his sorrow the fierceness of the Sioux, the "tigers of the plains," Verendrye determined upon a route farther to the north into the haunts of the Assinniboines and the Cristineaux, mortal foes of the savage Sioux, who hunted and trapped as far north as the country now known as Manitoba.

The starting-point of Verendrye's expedition was Montreal; the date, June 8, 1731; the men, Verendrye, his three sons, a nephew, La Jemeraye, and a number of Canadians; the object, the discovery of the Sea of the West, the aggrandizement of the French nation, and the winning of great wealth and fame. But Fate seemed to have turned her hand against him. His youngest son, his nephew, and many of the Canadians were killed by the Indians. The financial support that he had reason to expect was not given; supplies coming to him were lost or stolen on the way; and no definite information as to the desired route could be obtained. Finally, the Assinniboines and the Cristineaux told him of a tribe of Indians living on the Missouri River "many moons" distant, who could guide him to the much coveted sea. In 1738, October it was, Verendrye began his exploration to the west, arriving at the Mandan Villages three months afterwards. Here he found six villages of Indians.[1] On account

[1] When Lewis and Clark visited these villages in 1804, they had been by that time reduced to two villages and had been moved up the river about fifty-five miles from the original location. This migration was caused by the persecutions of the Sioux, who were the Mandans' mortal enemies, and smallpox, both of which had greatly reduced their numbers.

of the desertion of his interpreter and the loss of the luggage containing the trinkets indispensable for trading with the Indians, Verendrye was obliged to retrace his steps to Fort La Reine early in 1739. Ill health combined with the trials and exposure endured in this western trail made it impossible for the elder Verendrye to make another attempt to reach the mythical sea.

In the spring of 1742, his sons, Pierre and the Chevalier, with two Canadians, again visited the Mandan Villages. From here, in July, with two Mandans in addition, the party pushed to the west-

From Maximilian's Travels.
Courtesy of The Arthur H. Clark Co.
A MANDAN CHIEF, MAT-TOPE, IN FULL DRESS

southwest between the upper Missouri and the Black Hills. Game of all kinds was encountered, including elk, mountain sheep, antelope, deer, wolves, and the ever-present prairie-dog. Far west, it may have been as far west as the Yellowstone River, the Mandan interpreter deserted the French brothers, leaving them with an unknown tribe of Indians in an equally unknown country. From here a western direction was taken, in the course of which march they met a band of Little Foxes, who with other tribes were in mortal terror of the much hated Snakes, or Shoshones, who lived some distance toward the desired sea. Finally, after a journey of a few days to the southwest, they

From Maximilian's Travels
INTERIOR OF A MANDAN HUT

came to the Bow Indians, who knew of the coveted sea only through information given to them by captive Snakes. East, west, and east again they journeyed, until on New Year's day, 1743, they sighted the Big Horn Mountains, a branch of the Rockies, somewhere near the present Yellowstone National Park in the present state of Wyoming. They came within one hundred twenty miles of this museum of nature. From here they went south on Shoshone River, down to Wind River (in the central part of Wyoming), where the natives told them of Green River beyond the mountains. To the Verendryes belongs the credit of being the first white men to see the Rocky, or "Shining," Mountains. Here ended their journey westward. Turning their faces to the east, after many long weeks of travel they reached Montreal in May, 1744, having spent eleven years seeking to find the waters that Lewis and Clark reached more than a half-century afterward.

From Maximilian's Travels
A LITTLE FOX INDIAN, WAKUSASSE

Their journey was a failure in the same sense that Coronado's was. They did not find that for which they sought, but they were the pioneer explorers of the northwest as was Coronado of the southwest. They left a lost trail to be remade by others of another century. Had the Verendryes gone less than one hundred miles farther south they might have discovered South Pass, the great gateway in the path to the West at the end of which was the much desired "Sea of the West." Like Coronado, their dream of many years was not realized, and they faced defeat, obscurity, and poverty. These are some of the rewards that come to the pathbreakers whose dreams become a reality for the next generation. Somewhere, not far from the present southern boundary of South Dakota, lies buried in the soil and rocks a leaden plate bearing the arms and inscription of the King of France. This was placed there by the Verendryes to commemorate their expedition, a monument to French enterprise.

3. LEWIS AND CLARK

La Salle's dream was that there was a waterway from the region of the Great Lakes to the Pacific Ocean. Lewis and Clark made his dream a reality. Where the Verendryes failed, Lewis and Clark led a triumphant march. A land route was discovered to the Pacific, the river flowing to the sea definitely located, and the western boundary of the United States transferred from the Mississippi and Missouri rivers to the Pacific coast.

Louisiana, the purchase of which was made by the United States from Napoleon, was a vast, unbounded wilderness west of the Mississippi. This land first belonged to France by the right of discovery, through the exploration of the Jesuit Missionaries, Fathers Marquette and Hennepin, and especially through the efforts of the brave La Salle. In 1682 La Salle took possession of the land down the Mississippi to the Gulf of Mexico in the name of Louis XIV of France, and named it Louis-i-anne. Spain took possession of this territory in 1763, and the United States in 1795 entered into a very indefinite and unsatisfactory treaty with Spain in regard to the use of the mouth of the Mississippi, the site of the present New Orleans. Spain possessed both sides of the river at its mouth, and controlled the western shore to its source. The denial of the free use of the highway was a real and serious injury to the frontier people. The Americans asked only

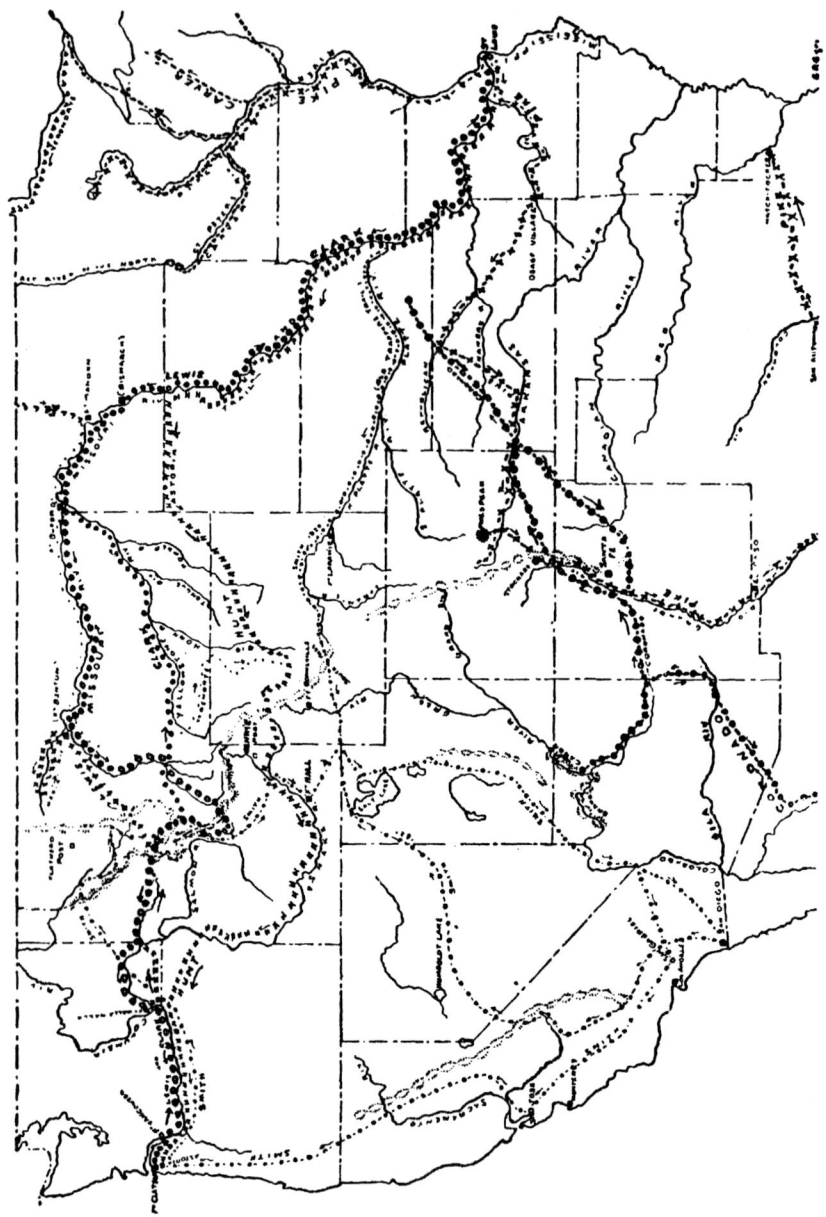

MAP I. THE PATHS OF THE EARLY EXPLORERS

⊤O−O−O	Coronado, 1540.	− − − − −	Carver, 1766–68
V V V V V	The Verendryes, 1742–44.	−×−×−×	Pike, 1805.
O O O O O	Lewis and Clark, 1804–06.	H H H H H	Hunt, 1810–12.
× × × × ×	Lewis to Maria's river.	−●−●−●−	Smith, 1826–29.
▬ ▬ ▬ ▬ ▬	Lewis and Clark cut off on return.		

that they might have free use of the river as a highway, and a small area at its mouth where they might deposit their produce preparatory to transshipment to the Atlantic and European ports. Spain's refusal to grant the request almost brought on war.

The treaty of 1795 was only a temporary arrangement that at its best was most uncertain. Rumors of war, of a desire to take the mouth of the Mississippi by force, of the discontent arising from conditions which hindered the growth and prosperity of all of those who were dependent upon the navigation of that river, caused the authorities at Washington much anxiety. The bold frontiersmen were demanding in return for their allegiance, protection and aid from the United States.

In 1802 Spain closed the mouth of the "Father of Waters" to our products, thus virtually stopping the navigation of the river by the citizens of the United States. President Jefferson and his administration tried to plan ways and means by which the difficulty might be overcome. The President asked Congress to appropriate $2,000,000 for the purchase of New Orleans and West Florida from France, which had just acquired them again, a purchase which would carry with it the right to navigate the entire length of the Mississippi. Robert R. Livingston, one of the five to draft the Declaration of Independence, was at this time our minister to France. James Monroe in the early spring of 1803 was sent to Paris as a special envoy to assist in the purchase of New Orleans. Napoleon Bonaparte, First Consul of the Republic of France, was about to engage in a war with England, and, seeing that his enemy with her command of the sea could take Louisiana from him, he determined to sell the whole province. Barbé Marbois, Minister of the Treasury of the Republic of France, was given charge of negotiating the sale. These four statesmen, two of whom had taken part in gaining our independence, two of whom were decidedly conspicuous in the dramatic movements of the French Revolution, perfected an agreement by which all of Louisiana was to be added to the United States. Monroe and Livingston had asked only for New Orleans and the mouth of the Mississippi. Napoleon with a majestic wave of his hand, pointing toward the west, said: "I renounce Louisiana. It is not only New Orleans that I will cede. It is the whole country without reserve." The price was $15,000,000. The treaty was signed April 30, 1803; Congress ratified this November 3, 1803; the purchase was made December 17, 1803, when Livingston exclaimed, "We have lived long, but this is the noblest work of our lives." We have since learned that there were in the purchase 1,037,735 square miles, or about 664,150,000 acres. We paid, therefore, only two and one half cents an acre for the Louisiana Purchase. It has been said that this act was by far the greatest work of our Government during the years between the adoption of the Constitution and the outbreak of the Civil War.

Technically, France did not occupy Louisiana at the time of the purchase. The transfer from Spain had never been made. France did not possess the province she was selling. The formality of surrender from Spain to France had to be accomplished before France could surrender the land to the United States. November 30, 1803, with proper ceremonies, the yellow and red flag of Spain was lowered at New Orleans and the keys of the Island turned over to the French Representative, who then raised the colors of France. Following this, on December 20, the Tricolor descended the same pole down which the Spanish colors had traveled twenty days before, and the Stars and Stripes ascended, denoting the end of the rule of France on American soil.

Several years before the appropriation was made by Congress for the purchase of New Orleans, Jefferson, while

Secretary of State in 1792, agitated the question of sending an exploring party to navigate the Missouri River to its source. He had a desire to extend commercial relations with the Indians, and to obtain for our country some of the riches of the region which were being monopolized by the traders from Canada. No one had the slightest conception of the vastness of the territory lying beyond the Mississippi. Robert Gray had sailed from Boston around the Cape to the Pacific in the ship Columbia, casting anchor in 1792 in the harbor of a river which he named Columbia. Many English and Yankee ships were at this early date gathering furs on the Pacific coast, and the region about Vancouver Island was well known; but the region between the Mississippi and the Pacific was utterly unknown save to a few daring trappers who had ascended the Missouri a thousand miles or so, and had set their traps in some of its tributaries. Jefferson at one time had made arrangements with a John Ledyard of Connecticut to explore the Northwest by traveling eastward through Siberia, shipping at Kamchatka for the Russian port Sitka, coming down the coast and so across country to the American settlements. Ledyard journeyed from Paris, through Germany, Sweden, and Northern Russia, even into Siberia. Here he was arrested and taken to Poland with threats of death if he again attempted the exploration. Time has proved that Russia was anxious to do what she prohibited Ledyard from doing.

Still another attempt did Jefferson make to secure the exploration of this region when he persuaded André Michaux, a French botanist, to attempt it. Michaux started with a party from the Atlantic coast for the west, but before he reached the Mississippi he was recalled by his own government. All this goes to show that Jefferson's views on this great West were larger than those of his contemporaries. Indeed, if anything were needed to convince us of his statesmanship, a survey of his activities in relation to our western domain would furnish proof in plenty.

Three months before the treaty was signed transferring the Louisiana territory, Jefferson had sent a confidential letter to Congress asking for an appropriation of $2,500 to be used for equipping an expedition to explore the country that the United States claimed by right of discovery by Captain Gray. Strange as it may seem to us now that we know the value of that country, it was with difficulty that this amount was obtained. Had Congress but known that the property covered by the purchase would some day contain a taxable wealth of over $7,000,000,000 the paltry sum might not only have been more quickly granted but have been considerably increased.

Jefferson keenly realized that the success of the expedition depended upon the men chosen to conduct it. He made no mistake here. His former private secretary, Captain Meriwether Lewis, was chosen to take chief command, with Captain William Clark of Virginia as second. ⟨ There were in the party chosen for the expedition fourteen soldiers from the United States army, nine young men from Kentucky, all expert riflemen, two French watermen, one interpreter and hunter (Drewyer),[1] Clark's black servant (York), who proved to be a rare curiosity to the natives, and sixteen men who were to go only part way. Among the twenty men who were to take the entire journey there was not a married man. The instructions which Lewis and Clark received from President Jefferson were minute and complete. They were expected not only to observe carefully and keep full records, but also to be diplomatists in their dealings with the Indians, naturalists, botanists, geologists, paleontologists, astronomers, engineers, meteorologists, mineralogists, doctors, ethnologists, and, above all, ambassadors, for at all times

[1] The proper spelling is Drouillard, but pronounced as indicated.

they officially represented their country. Their journals are full of valuable descriptions of the various Indian tribes, many of them now extinct.

The expedition started from Wood River, opposite St. Louis, May 14, 1804, with three boats and two horses. The largest boat was fifty-five feet long, drawing three feet of water, with one large square-shaped sail and twenty-two oars; the other two boats were of six and seven oars; the horses were to go along the shore to help the boats when possible, and to bring to the boats any game that the hunters might shoot. The large boat had a "swivel" gun or small cannon swinging on a pivot which often came into efficient service, to make a loud noise if for nothing else. What with tortuous stream, unknown channel, numberless snags and sandbars, and swift current, progress was so slow that they counted themselves fortunate to make fifteen miles a day. It took them one hundred and sixty-five days to reach the Mandan villages, sixteen hundred miles from St. Louis. Coming home, down the river, they made forty-three miles a day, going over that part of the return voyage in thirty-seven days.

The bluffs, hills, creeks, rivers, were all named as they were discovered, the name suiting the object or chosen for some incident occurring at that time. Thus we have Bear and Antelope creeks, where these animals were first killed; Independence Creek, named on the Fourth of July; Floyd's Bluff, where one of their number was buried. We must remember that Lewis and Clark had no charts or maps to follow; it was all unnamed, unexplored territory to the white man. Their supplies for a trip of this magnitude were necessarily large. They had to take food, clothing, camp equipment, fire-arms and ammunition, in addition to innumerable articles for barter with the Indians they expected to encounter. Their food was to be the fish and game captured from day to day. Their powder was placed in numerous small packages, or canisters with lead wrappings. These wrappings served a double purpose, keeping the powder dry and furnishing just enough lead to make bullets for the amount of powder that was in the canister. In this way there was no waste of weight, a matter that had received the most careful consideration of both captains. Their supplies were put in several bales, each bale containing some of each article taken. Thus in case of accident or the loss of a single bale the entire supply of any one commodity would not be destroyed. In addition to these, there were fourteen other bales carrying presents for the Indians, consisting of bright colored beads, many blue ones, for they were the chief's bead and could be bartered for more than the other colored ones, tinsel and red cloth, laced coats, handkerchiefs of fancy colors, flags, medals, knives, toma-hawks, articles of dress, — anything to please the fancy of the bartering Indians. Lewis and Clark took three sizes of medals, representing as many grades of honor, that were to be given to the chiefs of the tribes they might encounter. A number of diaries or journals were kept, from which we have been able to obtain the most minute details of the entire journey, those of Lewis and Clark, of course, being the most complete and valuable.

There was not much variation from day to day in their experiences, the journey being rather uneventful for several months. They frequently met crude boats coming *down* the river loaded to the edge with hides and pelts for sale at St. Louis, — the beginnings of the fur trade along the Missouri River.

On August 2, 1804, they held their first formal council with the Indians, at which time Lewis told the chiefs of the changed government, made them promises of protection, and gave advice as to their future conduct. The chiefs were rejoiced at the change of government, and sent their regards to their "Great Father," the President. The spot where this council was held was called "Council Bluffs," whence the name of the present Iowa city.[1] Sioux City also is an historic spot, for it was here that Sergeant Charles Floyd, the only man lost during the expedition, was buried, and his grave is now marked by a handsome column.

From Maximilian's Travels
A MANDAN VILLAGE, NATIVES IN BOATS MADE FROM BUFFALO
SKINS STRETCHED ON A FRAME

Occasionally new kinds of animal life were seen to which names were readily applied. When the explorers saw their first prairie-dog, they named it "petit chien" (little dog), and they called the antelopes "goats."

On October 26 the explorers reached the Mandan villages, not the old ones known to the Verendryes, but the new ones, five days' journey farther up the river. Here they spent the winter, near the present Bismarck, North Dakota, housing

[1] The original Council Bluffs was on the west bank of the river, about twenty miles north of the present city.

themselves in huts and stockades and passing the winter in making boats, mending clothes, jerking meat, and learning the habits and customs of the Indians. A new interpreter was secured in the person of Toussaint Charbonneau, a French Canadian trapper who had worked for the Hudson's Bay Company. As he had worked for these fur men of the North he was familiar with just those things that Lewis and Clark did not know. Charbonneau took with him his young Indian wife, Sacajawea, and her papoose, an infant only a few weeks old. Since they expected to meet the Snakes, or Shoshones, it was thought that Sacajawea would be a useful additional interpreter, as she had been captured from that tribe when she was a child by the Minnetarees, by whom she had been sold into slavery to Charbonneau, who brought her up and afterwards married her.

At five o'clock in the afternoon of April 5, 1805, two expeditions left the Mandans, one to return to St. Louis with letters to the President, and with hides, stuffed animals, bones, articles of Indian dress, bows and arrows; the other, with thirty-two members in six canoes, to continue toward the unknown Northwest. Up the river they went, encountering plenty of deer, buffalo, elk, geese, ducks and prairie-chickens for food, and more bears than they found convenient or comfortable. As a matter of fact the bears became very dangerous, often questioning the right of man to infringe upon their domain. These encounters were important enough to cause a stream to be named "Brown Bear-defeated Creek."

Just north of Fort Benton is Maria's River. Where this stream joins the Missouri the explorers were in a quandary as to which stream was the branch and which the main river. Lewis, after spending four days on the northern branch, or the Maria's, finally decided that the southern one was the one desired and returned to the junction. At the meeting of these two forks, in order to lighten their loads and also to have supplies when they returned from the West, they *cached* many things.[1]

Still burdened to the limit of their strength, they set out on foot to find the Great Falls of the Missouri of which the Indians had told them. Lewis discovered these tremendous falls on the 15th of June, 1805; and as he stood watching the mad rush of waters his thoughts flashed forward to the time when a great city would grow up about this storehouse of power. Thus felt Champlain upon first beholding Niagara, and Father Hennepin when he saw the Falls of St. Anthony. This is the greatest recompense of the explorer — to be first to find a wonder so full of splendid possibilities for the future. To get supplies around the falls it was necessary to *portage* past them, a task occupying two weeks. Many of the things were carried on the men's backs, but most of them in a cart, the wheels of which were made from a cotton-wood tree two feet in diameter. Moccasins were the only covering for the men's feet, and so poor protection did they afford that the prickly pear, or cactus, easily pierced them and left the feet raw and bleeding. Cactus, heat, fatigue, and hard work were a fearful strain on the men. But, the

[1] This is a most common method adopted by the mountaineers to take care of the things. A good, dry spot is selected; the sod is carefully removed and placed to one side, so that when it is replaced it will not show that it has been disturbed. After the sod is removed a hole is dug, and the extra earth that will not be needed to fill up the hole is carried to a stream and thrown into the water, so that no trace of it may be seen. Then twigs and branches are placed in the bottom of the hole, and on these are placed the goods to be cached or hidden; then these are covered with hides and skins to keep out moisture or water; over all of this is placed enough of the dirt to fill the hole, leaving space enough for the sod, which is carefully replaced. Sometimes a fire is made on the spot to destroy any sign of the work, or horses are picketed over the cache. If the greatest care was exercised, even the skilled eye of the Indian could not detect the hiding-place.

portage finished, once more they embarked and soon were at the three forks of the Missouri, near the present town of Three Forks near Bozeman, Montana. They named these parent streams the Jefferson, the Madison, and the Gallatin, after the three statesmen who were then guiding the affairs of this country. Sacajawea recognized this to be the exact

Northern Pacific Railway
THE THREE FORKS OF THE MISSOURI, NAMED BY LEWIS AND CLARK, JEFFERSON, MADISON, AND GALLATIN

spot where she had been captured five years before. When the explorers arrived at this point in their journey they were two thousand eight hundred forty miles from home. Lewis and Clark felt at this stage of their travels that they were approaching Sacajawea's country, and at any time might encounter the hostile Indians. Every precaution was taken, as the entire success of the journey depended upon the friendly relations that might be established with these Indians. Sacajawea, like a homing pigeon, knew the way, guiding here, directing there, pointing to this and that.

Occasional signs were observed of Indians; a wild horse or a worn moccasin, or smoke signals, all indicating that an experience was soon to be theirs.

It now became absolutely necessary to find mountain Indians to furnish the party with horses to cross the mountains, and with guides as well. Captain Lewis, going ahead of the rest with two of his men, discovered a man riding on horseback. Although the Captain made the friendly sign usual with the Missouri River and Rocky Mountain Indians, of holding his blanket in both hands at the two corners, throwing it above his head and unfolding it as he brought it to the ground as if spreading it out on the ground and inviting them to come and sit on it, he failed to attract the Indian. Then he ran toward the Indian with a looking-glass and trinkets calling, "tabba bona," "tabba bona," words that Sacajawea had taught him to use, meaning "white man," "white man," at the same time rolling up his sleeves and opening his shirt to show the white skin of his arms and chest, for his face and hands were browned and tanned to the color of an Indian from the exposure to sun and wind during months of outdoor life. But the Indian fled. The next day, however, Lewis overtook some squaws who conducted him to a camp where he met a chief and about sixty warriors, all well mounted. The chief, Cameahwait, after much bartering and bickering, agreed to furnish horses and a guide to pilot the expedition over the mountains.

Early the next morning Clark, Charbonneau, and Sacajawea, with the rest of the party came into the camp. Just before their approach Sacajawea commenced to jump and dance with joy, sucking her fingers, a sign of joy with her tribe. Suddenly she threw her arms around the neck of an Indian woman, crying and laughing. This was her lost companion who had escaped and returned home when Sacajawea had been taken captive. When Chief Cameahwait appeared she rushed to embrace him, throwing her blanket over their heads, weeping and showing the most extravagant joy. They were brother and sister. From him she learned that her sister had died since the time of the tribal battle of which we have spoken, the sole representative of the sister's family being a small boy whom Sacajawea immediately adopted. History does not record what was done with the boy while the expedition journeyed on farther to the West, but the strong presumption is that he remained with the expedition, went to the coast with his adopted mother, and finally went to the Mandan villages. It may be well to keep this boy in mind, for we shall learn more of him after he grew to manhood.

At this point there must be some expression of admiration for this little Indian woman who during the night after her arrival in the camp heard, while in her tepee, her brother and his men plotting to drive away to the mountains the horses that Lewis had purchased and to leave the expedition without any possible means of getting on, as their baggage was too heavy to be carried on the men's backs. In the morning she told Clark of the proposed dishonesty, thereby casting her lot with the white men rather than with her own tribe. The white people had been kind to her, and she felt an obligation and desire to be loyal to their cause. There were many other times up to this point in the journey when Sacajawea had rendered valuable service, really service that was invaluable. When the expedition was at Brown Bear-defeated Creek the boat turned over and the valuable papers, some scientific instruments, medicine, and almost every indispensable article for the journey spilled into the water. They were rescued by the quick and intrepid action of this Indian woman. Without these scientific instruments it would have been useless to proceed. To have returned to civilization to obtain new ones would have postponed the journey for one year at least, if not indefinitely. Lewis and Clark named one of the streams Sacajawea, but it is known to-day as "Crooked Creek."

It was Sacajawea who found the pass in the mountain for Clark on the return journey between the Gallatin and Yellowstone rivers. This is now known as the Bozeman Pass and is located between the Bridger and Gallatin ranges, east of Bozeman, Montana. Charbonneau was the interpreter of the expedition, but Sacajawea often had to come to his rescue in this work. One interesting circumstance will illustrate how hard it was to hold a conversation. There was a controversy over some horses at a time when the possession of horses meant success or failure. The contestants were of the Chopunnish tribe. One of Lewis and Clark's men took the wording of the trial in English and turned the English into French for Charbonneau, who translated this French into Hidatsa for Sacajawea, who gave the Hidatsa in Shoshone to the Shoshone Indians, who in turn adapted this Shoshone to Chopunnish for the contesting Indian chiefs.

Service in a medical way was often distressingly needed. In this particular Sacajawea added to her usefulness, for her native knowledge of medicinal herbs and of the curative properties of plants was of extreme worth in time of sickness. Again it is difficult to imagine, when starvation seemed to be the only outcome, what would have been the result if she had not concocted messes made from seeds and plants, and had not known of the treasures of artichokes stored away in the prairie-dog holes.

In September of this year, 1805, the party crossed the Bitter Root Mountains amidst snow and drifts. Here they hailed with delight the first westward-flowing streams. Would they empty into the Pacific? That was the question. The mountains crossed, they left the horses, after branding them,[1] in care of a Nez Perce chief, built canoes and again embarked, floating with the stream now instead of toiling against it. This stream, the Clearwater, emptied into the Snake where Lewiston, Idaho, and Clarkston, Washington, now stand. Down the swift-flowing Snake they sped to its junction with the Columbia, a short way above the present Kennewick and Pasco, Washington. Embarked on the broad Columbia, their course was rapid and easy, save for the excitement of shooting an occasional rapid or making a portage. Soon Mt. Hood was seen to the south, Mt. Adams to the north. On the east side of the mountains game had been very abundant, and the explorers had plenty to eat. On the west side they actually suffered for proper food. In fact they became so reduced for necessities that they were obliged to buy puppies from the Indians, with which they made stews that were very much relished. On October 28 Indians visited them, one of whom had on a round hat and a sailor's peajacket; another one had a British musket; one, a cutlass; one, a brass teakettle; one, bright colored cloth. Then it was that Lewis and Clark knew that they were near the end of their journey, as these things must have been obtained from traders who had reached the shores of the Pacific by the water route. The roar of the ocean was a thrice welcome sound to their ears, and on November 8 their eyes were rested and their hearts rejoiced by the sight of the goal of their many months of toil and travel, — the Western Sea, the goal of the Verendryes' Expeditions. One week later Lewis and his party reached the ocean and built Fort Clatsop on the south side of the mouth of the Columbia, four thousand one hundred thirty-four miles from home. Here they established their winter quarters, and stayed until

[1] This branding iron was found in 1892 on one of the islands of the Columbia, three and a half miles above The Dalles, and is kept as an interesting relic in the possession of the Oregon Historical Society.

March 23 the next year, 1806. The men were almost naked. Clothes had to be made for immediate wear and for the return journey. During the winter Captain Clark made a map of the entire country over which the expedition had traveled, and the men made over four hundred pairs of moccasins, many gallons of salt from sea water evaporated, numerous packs of jerked venison, and clothes of all kinds from the skins of elk, deer, beaver, and sea otter.

Many tribes of Indians were visitors to this fort during that winter. Comcomly, the one-eyed chief, was mentioned as frequenting the fort. We hear of this old chief again when he enters into the life of the Astorians who wintered in this exact locality.

Captain Clark's drawing.   Dodd, Mead & Co.
CLATSOP, OR FLATHEAD, INDIANS, SHOWING METHOD OF SHAPING FOREHEAD

Lewis and Clark were in hopes that any day a vessel might arrive from the States by the way of the Cape or from China. Before leaving they posted on a tree a notice telling of their trip, and with it left a map of the route taken. Subsequently, this was found and given by the

Northern Pacific Railway
POMPEY'S PILLAR, YELLOWSTONE VALLEY, MONTANA, FIRST MADE KNOWN AND NAMED BY CAPTAIN WILLIAM CLARK IN 1806

natives to a Captain Hill, who was in command of a ship that arrived at the mouth of the Columbia too late to be of service to Lewis and Clark. Captain Hill took it to China, and brought it thence to the United States. On March 23, 1806, after presenting Chief Coboway with Fort Clatsop and the furniture, they pushed the boats from the shore and the home journey began.

The return journey was easier and more quickly made. In one place alone the expedition saved eighty miles by a short cut in the southeastern part of what is now the state of Washington. When they reached the Clearwater, they found that the Nez Perces had been true to their trust, for the horses were there and were promptly delivered to their owners. It was at this camp that they had to resort to medical service to obtain necessary supplies. Eyewater was at a premium, cures for rheumatism in great demand, and Lewis and Clark rendered to the suffering Indians helpful service, thus obtaining their good will. When the expedition left the coast they had on hand for exchange only "six blue robes, one of scarlet, five made out of the old United States' flag that had floated over many a council, a few old clothes, Clark's uniform coat and hat, and a few little trinkets that might be tied in a handkerchief." Finally the medicine chest was empty; brass buttons had departed from the soldiers' uniforms; needles, thread, fishhooks, files, and awls had been exchanged for bread, and now these "made-to-order" doctors had to resort to every device to obtain from the natives the commonest commodities. In the eyes of the Indians they were wonderful medicine men. One can fancy how mystified the red men were over a watch, with its even ticking inside of its covers; the magnet, with its power to make a piece of steel move without touching it; phosphorus, invisible in the daytime but ghastly at night; the spyglass, bringing objects at a distance within the reach of the hand;

the burning glass, stealing fire from Heaven; and the air gun, with its terrifying noise. No wonder the Indians thought the white men were more than human beings. The exhibition of these mysteries brought to the party the much needed food and supplies.

After they had crossed the Bitter Root Mountains, the party divided, Lewis going north by the Missouri River, Clark south, by the Yellowstone. Captain Lewis was not satisfied with his previous exploration up Maria's River. Taking with him Drewyer, the interpreter and hunter, and the two Field brothers, he traveled northeast to what is now Great Falls, Montana, and then northwest, trying to find the source of the Maria's River. The Blackfeet, the most treacherous of all Indians, were encountered. In the camp one early morning an Indian stole Field's rifle. Field regained this after stabbing the thief to the heart. In the heat of the struggle the Indians were discovered trying to drive away the horses. Without horses, Lewis and his men would have been helpless, miles from their party, in a strange country, surrounded by hostile Indians. What was to be done had to be done quickly and effectively. Lewis at once shot and killed a Blackfoot, in exchange receiving a shot that fortunately only passed through his hair, seized the horses of the braves, and the whole party fled for the Missouri. They rode sixty miles without stopping, covering over one hundred and twenty miles in twenty-four hours. Opening their *caches* at the junction of the Missouri and the Maria's rivers, they found much of their buried goods spoiled by water, but the iron boat was in good condition. Turning the horses loose on the prairies, they went down the Missouri in the boat and met Clark, Charbonneau, Sacajawea, and the rest of the party at the junction of the Yellowstone and the Missouri.

On the way to this junction, Lewis, while hunting one day, was mistaken for an antelope by one of his men and was shot in the hip. While the injury was not serious, when he met the rest of the party he turned over the command of the expedition to Clark, who had successfully made the journey down the Yellowstone and had arrived before Lewis and his division. When the boat arrived Clark was

YELLOWSTONE NATIONAL PARK. ANGEL TERRACES AT THE
MAMMOTH HOT SPRINGS

John Colter was the first white man to see this formation. When he told of the wonders in this location the people of St. Louis thought that he had lived too long alone in the wilderness and had become mentally unbalanced.

much disturbed at not seeing Lewis, and his joy at finding him alive, though wounded, testified strongly to the love these two fine young men felt for each other. Two days after this the entire expedition arrived at the Mandan village, where they found that during their absence their fort had been destroyed by fire.

Charbonneau, Sacajawea, and the baby did not continue with the expedition but stayed at the Mandan village. The interpreter was paid $500.33 for his services, and Clark expressed his appreciation of the services rendered by the little Indian woman. One man, John Colter, at his own request was permitted to return to the wilderness to trap. While trapping he won for himself the distinction of being the first white man to witness the thrilling wonders of Yellowstone Park. From Mandan, Lewis and Clark took Chief Big White with them to St. Louis and ultimately to

YELLOWSTONE NATIONAL PARK. THE GIANT PLAYING
This geyser action takes place periodically from four to seven days. John Colter, of the Lewis and Clark expedition, saw these wonders.

Washington, D. C. The remainder of the voyage was an easy matter, down the stream. They arrived at St. Louis September 23, 1806. Great was the rejoicing at this successful termination of the unparalleled undertaking, for the people of that city looked upon the explorers as if they had risen from the dead, so long a time had elapsed since information about them had been obtained. Verendrye had failed; Lewis and Clark had succeeded.

Captain Lewis, was appointed Governor of Louisiana Territory, which office he held until his death, two years later. Captain Clark was appointed General of the militia and Indian Agent of the territory. He administrated affairs for the Indians with rare judgment, being fondly called "Red Head" by the Indians, while the city of St. Louis was known as "Red Head's Town."

Very recent investigations have brought to light an interesting page in history relative to Sacajawea.[1] Practically all trace of the Indian woman had been lost for a century. Meager mention had been made of her by writers of history who repeated information obtained from traders and trappers along the Missouri. But in the endeavor to obtain a Shoshone woman to serve as a model for a statue representing Sacajawea, in commemoration of the Louisiana Purchase, valuable data were unearthed which proved that Sacajawea was buried on the Wind River, or Shoshone, Reservation in Fremont County, Wyoming. This information showed that she had died at an advanced age, in 1834, and had received a Christian burial, through the efforts of the Episcopal missionary to whose church she belonged.[2] The son, Baptiste, she had taken to the coast on her back and the nephew, Bazil, whom she had adopted when on her way to the Pacific coast, had lived with her for years in this fertile valley, where they in turn had been pilots and scouts for the white man in his travels and his hunting expeditions through that region. It is a custom of the Indians that after they once adopt a child they never admit the adoption but give to him the position he would have occupied had he really been a son. Thus Sacajawea's adopted son, who was older than her own boy, was always

[1] "Pilot of the First White Men to Cross America." The Journal of American History, Vol. I, No. 3. Grace Raymond Hebard.

[2] In the parish records appears: "April 9, 1884. Bazil's mother, Shoshone, one hundred years, residence Shoshone Agency. Cause of death, old age. Place of burial, Burial Grounds, Shoshone Agency."

the one that took care of Sacajawea. He wore on his neck
at his burial a medal that he claimed his father Charbonneau
had given him, as Charbonneau, himself, had received it
from Lewis and Clark after the expedition returned to
Mandan.

4. ZEBULON PIKE

August 9, 1805, Captain Lewis was pausing on the Continental Divide, near the headwaters of the Jefferson,
preparing to cross to the Pacific slope. On this exact date

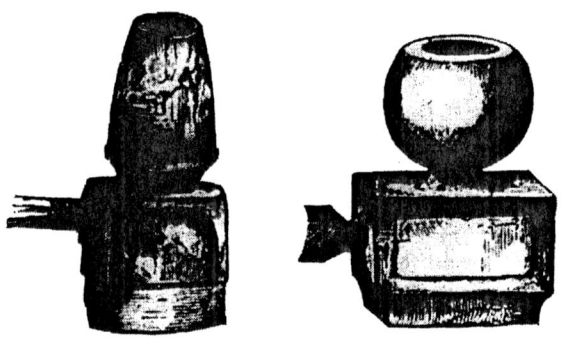

INDIAN PIPES MADE OF TALC

another indomitable explorer whose name will always be
inseparably linked with those of Lewis and Clark started
from St. Louis by the way of the Mississippi, wishing to
locate the headwaters of that river, and to ascertain the
extent and value of the newly acquired territory embraced
in the Louisiana Purchase.

President Jefferson was anxious to justify his purchase of
this wilderness, so he sent Zebulon Pike up the Mississippi
as he had sent Lewis and Clark up the Missouri. Pike was
put in command of the expedition by General Wilkinson, with
orders to explore and make a report upon the Mississippi
to its source, to make peace with warring Indians, particularly the Ojibways and the Sioux, to select desirable sites

for military posts, and to ascertain to what extent the British fur traders were still occupying our territory recently purchased from Napoleon.

This country into which Pike was now to travel had been explored by Jonathan Carver in 1766–68, when he was hunting for the headwaters of the Mississippi. He had

Northern Pacific Railway
A BLOCK-HOUSE, FORT RIPLEY, ERECTED IN THE REGION OF PIKE'S EXPLORATIONS

fully described that country to the north and northwest of the head of Lake Superior. The French were familiar with stories similar to the ones he told about the Indian tribes living to the west in the "Shining Mountains" where gold was in such abundance that the most common utensils were made from it. These exciting tales spurred Carver more than once to try to cross the continent, but always without success.

Without particular incident or accident, Pike with his twenty men navigated as far north on the "Father of Waters"

Northern Pacific Railway

SAINT PAUL IN 1851, EIGHTY-FIVE YEARS AFTER CARVER CAMPED ON THE SITE

as Little Falls, Minnesota. Here he left some of his men in
a stockade which they had built, and pushed up overland to
the mouth of Turtle River into the regions explored by
Jonathan Carver in 1766–68. This was as far as Pike
attempted to advance. He found British fur men in the
country and protested to them, saying that they were now
in the country owned by the United States.

Northern Pacific Railway
SAINT ANTHONY FALLS, MINNEAPOLIS
Viewed by Carver and Pike

Pike returned to St. Louis in April, 1806. From here,
with twenty soldiers, he started on his second expedition in
July of this year, going westward into Louisiana Territory.
New Spain, or Mexico, was jealous of the possessions the
United States had acquired, and was ready to contest every
mile that our government might attempt to claim. We
must remember that the exact boundary lines of the Louisiana Purchase were not defined. When an Emperor deeds
a territory to a nation fixing its limits by the wave of his

hand the boundary lines are liable to be uncertain. This second journey of Pike's had for its main object the discovery of the course of the Arkansas River and the location of its headwaters. This and the Rio Grande were felt to be determining streams in settling any boundary between the United States and Spain. Jefferson felt that we must have some definite knowledge of that southwest region in case of dispute with Spain and he sent Pike to get this knowledge.

Pike went farther and learned more than any one had hoped or even wished. His purpose was to go up the Arkansas until he came to the mountains and then to go south to the Red River, returning home on that stream. After traveling many days, weeks in reality, Pike discovered, November 15, 1806, a mountain, which looked to the naked eye like a small blue cloud. A half-hour's travel brought him in full view of the peak which now bears his name. His party with one accord "gave three cheers to the Mexican Mountains," which shows clearly that they did not know where they really were. Confident that the lofty peak could be reached in a few hours they pushed on, shivering, cold and poorly clad. By this time Pike and his men were without shoes, using skins to cover their feet. Their thin summer clothing was worn to rags. After marching for twenty-five miles they found themselves at evening apparently no nearer the mountain than they had been at sunrise. Pike attempted to climb the peak but was obliged to abandon the attempt, declaring that it would be impossible for man ever to reach the top. Not only has man reached the top but a train daily takes scores of human beings to its summit. Pike's Peak will ever be nature's monument to this bold explorer.

Notwithstanding the heavy storms and the lack of clothing Pike went up the Arkansas almost to its source. Then he led his band to the southwest searching for the headwaters of the Red River. The cold became intense. Some of his party were so badly frozen that they were crippled for life. They could go no farther. Building a cabin to shelter them and leaving them nearly all his provisions, Pike, with the few men still able to march, hurried southward, seeking relief. Great was their joy when they found themselves on

ZEBULON MONTGOMERY PIKE   (Colorado Springs)

a stream flowing southward. Now they were sure that they had found the Red River. Here they built a stockade, and one of the party, Dr. Robinson, went directly to Santa Fé for succor.

Pike was in a serious predicament, for the stream was not the Red River at all, but the Rio Grande. He had invaded Spanish soil, and here he was captured by Spanish dragoons under orders of Governor Allencaster who had heard that armed forces, carrying the American flag, were on his territory. The arrest or capture was really in the form of a polite invitation to visit the Governor at Santa Fé. On the 6th of February, 1807, after a breakfast on deer, meal, goose, and biscuit, which an Indian spy had brought to Pike and his men, the commanding officer of the dragoons said: "Sir, the Governor of New Mexico, being informed that you had missed your route, ordered me to offer you, in his name, horses, mules, money, or whatever you might stand in need of to conduct you to the head of Red River, as from Santa Fé to where it is sometimes navigable is eight days' journey, and we have guides and the routes of the traders to guide us."

"What," said Pike, interrupting him, "is not this the Red River?"

"No, sir; the Rio del Norte."

At this Pike immediately ordered the United States flag to be taken down and rolled up, feeling that he had committed himself in entering Spanish territory, and that they had justifiable reasons for his arrest. Upon his consent to accompany the Spanish soldiers to Santa Fé, he and his men were supplied with clothing and food. Pike was treated with every consideration, notwithstanding that he had invaded Spanish soil. When he arrived at Santa Fé he was ordered to explain how he came to be with an armed force within the territory of the Spanish. Evidently he could give no excuse that seemed good to his Spanish captors, for they kept him and his men prisoners in Santa Fé for several months and then marched them to Mexico under strict guard. After conducting the prisoners through various parts of New Mexico and Chihuahua, always keeping them

SANTA FÉ TO-DAY

under strict surveillance, but treating them with consideration, the Spaniards brought them through Texas and turned them loose at Natchitoches on the Red River. Pike had not succeeded in preserving all of his notes,[1] but he had used his *eyes* and his ears well, had learned much about that strange Spanish land, its riches, its love of finery, its dependence upon old Spain for supplies, and to his report we can

[1] Many of his records were stolen by the Spanish authorities, but Pike, suspicious that his precious notes might be confiscated, concealed his smaller note-books in the barrels of the guns of his men.

trace directly the beginning of the Santa Fé trade that began soon after and proved so rich a field for American enterprise.

REFERENCES

Prince.   Historical Sketches of New Mexico.
Johnson.   Pioneer Spaniards in North America.
Winship.   The Journey of Coronado.
Thwaites.   Rocky Mountain Explorations.
Dellenbaugh.   Breaking the Wilderness.
Bancroft.   History of Mexico, Vol. II.
Parkman.   A Half Century of Conflict.
Laut.   Pathfinders of the West.
McMurray.   Pioneers of the Rocky Mountains and West.
Hebard.   History and Government of Wyoming.
Journals of Lewis and Clark.
Wheeler.   The Trail of Lewis and Clark.
Hosmer.   The Expedition of Lewis and Clark.
Hermann.   The Louisiana Purchase.
Schafer.   History of the Pacific Northwest.
Dye.   The Conquest.
Coues.   The Expedition of Zebulon Pike.
Hough.   The Way to the West.
Inman.   The Old Santa Fé Trail.
Bennett.   A Volunteer with Pike.
Grinnell.   Trails of the Pathfinders.

CHAPTER II

THE FUR TRADERS

1. THE MISSOURI RIVER MEN

When Lewis and Clark came down the Missouri in the fall of 1806, they met many venturous traders in furs wending their way to the great country in which the explorers had passed the last two and a half years. Reports of their successful journey up the Missouri as far as Fort Mandan had reached St. Louis long before Lewis and Clark had returned. Their reports, telling of many fine game regions and good beaver streams in the far western mountains, gave a tremendous impetus to the trapping industry. Yet one must remember that fur traders had penetrated the country of the northwest, at least as far as the Mandan Villages, how far is not known, before Lewis and Clark started on their expedition. Witness the boats loaded to the brim with furs coming *down* while the explorers in 1804 were going *up* the Missouri.

St. Louis was the general starting-point for the expeditions that went up the Missouri for the Northwest. Here the outfits were equipped for the long voyage, and thither they returned with their boats filled with precious hides from the north. These furs and peltries were obtained from the beaver, fox, otter and mink, and the heavier skins from the buffalo. The skins of the buffalo were used for robes and winter coats. Beaver skins were in great demand for making beaver hats, so much worn by men of fashion.

In the spring of 1807, Manuel Lisa, who had organized the Missouri Fur Company, with himself as its head and Captain Clark as its agent in St. Louis, went north and built Fort Manuel at the mouth of the Big Horn River. Here he planted himself, sent his hired trappers and hunters out to gather furs, and invited the Indians in to trade their furs for beads, flashy cloth, knives, hatchets, tobacco, and, it must be admitted, guns, powder, lead, and poor whiskey. The guns were not very good ones and did no one much harm, but the whiskey was especially bad and worked havoc. This bringing of whiskey to the Indians was the greatest sin of the trapper. The Government later worked faithfully to stop the practice, but it was hard to kill. Early in the spring of 1810 the Missouri Fur Company took a long stride into the wilderness by sending Andrew Henry with a strong party to the Three Forks of the Missouri. Three hundred packs of beaver skins were taken at the Three Forks during the first year. But the Blackfeet swooped down on the fur men, killing five of the trappers and taking their horses, traps, guns, ammunition, and all of their furs. Other attacks followed, in one of which George Drewyer, the hunter for Lewis and Clark, was killed after a desperate defense. These reverses so discouraged the company that Henry abandoned his post, crossed the Continental Divide to the south, and established himself on the north fork of the Snake River which is known to this day as Henry's Fork. This tireless trapper then built a temporary post near what is now the little town of Egin, Idaho. Game proved to be very scarce in this new locality, and in the spring of 1811 Henry returned down the river, meeting Lisa en route. It was on this trip up the river to relieve Henry that Lisa had the exciting race with Wilson Price Hunt, one of the

Astoria men on his way to the Pacific. Lisa gave invaluable aid during the war of 1812. He was made a sub-agent to all of the Missouri tribes above Kansas, as he was the man who best understood the Indians. It was through his efforts that all of these tribes along the upper Missouri allied themselves with the United States, rather than with the British fur men, and re-

INDIAN DRINKING-CUP MADE OF HORN

mained at peace when they might have desolated the border. Lisa died in 1820, after having traveled up and down the Missouri at least a dozen times, or a distance of twenty-six thousand miles by water. "He was beyond comparison the ablest of the traders so far as the actual conduct of an enterprise was concerned, and wherever he alone had control, and when not hampered by the counsels of others, he generally succeeded."[1] One might say that Manuel Lisa was the Cortez of the Rocky Mountain trade.

2. ASTORIA

One of the first men to grasp the tremendous advantages offered for a transcontinental route, in the region along the Missouri River and down the Columbia, was John Jacob Astor, the founder of the fur post, Astoria, at the mouth of the Columbia River. The reports of Lewis and Clark showed that this new country abounded in furs, which might be carried to the mouth of the Columbia and from there be sent to China, where there was a great market for them. "With China a market for furs from the Pacific coast, with Russian establishments on the northwest coast, which his

[1] Chittenden. History of the American Fur Trade.

ships might supply as an incident to their main business, with markets at home for the products of the Orient, with lines of trading-posts all along the Columbia from the sea to its source, connected thence with the Missouri and extending down that stream to St. Louis and from that point by the way of the Great Lakes to New York itself, Mr. Astor saw that his business would indeed be world-wide in scope and international in importance."[1]

John Jacob Astor was born in the German village of Waldorf, near Heidelberg, on the banks of the Rhine. When he was a youth he had a dream of great wealth that he was to acquire. Leaving home he went to London, thence to New York, where he had an elder brother. When he came to New York in 1784 he brought with him certain merchandise which he exchanged for peltries. These he sold in London at a great profit, convincing himself that there was much money to be made in the fur business. The United States at that time did not have any organized fur business, so the young Astor made annual trips to Montreal to purchase furs. As no direct trade between the United States and Canada was at that time allowed, these furs had to be shipped to London. When these restrictions of trade were removed he sent the furs directly to New York, and from there shipped them all over Europe and even to China, bringing in exchange dress goods, jewelry, tea, coffee, and spices to the United States.

Astor established the Southwest Fur Company, exploiting the region of the Great Lakes and the upper Mississippi; but soon his plans became world-wide in scope. He planned to establish posts all along the Missouri and the Columbia to the Pacific. Then, on all of the streams tributary to these rivers, were to be built smaller posts, trading directly with the Indians for furs. The furs along the northwest coast

[1] Chittenden.

were to be gathered by his ships, and the Russian Fur Company post at Sitka would get its supplies from him. Supplies for all these posts were to be sent to the western coast by ships from the Atlantic.

That he might have experienced lieutenants in this great undertaking, Astor chose for his partners, with one exception, factors of the Northwest Fur Company, a Canadian concern, then in fierce rivalry with the Hudson's Bay Company. Even his clerks came from the ranks of the enemy. None of these men put any of their own capital into the enterprise. Astor furnished all the money. Wilson Price Hunt of New Jersey was the sole American partner. The names of the others read like the roster of the Scottish Clans: Alexander McKay, Duncan McDougal, Donald McKenzie, Robert McLellan, Robert Stuart. This selection of partners was Astor's serious blunder, not because they were foreigners, but because they had a foreign interest. They had really no interest, pecuniary or patriotic, to lead them to fight for the success of the Pacific Fur Company.

Two expeditions departed from our eastern shores for the mouth of the Columbia River, one by land and one by sea, one to go around Cape Horn, the other over the route of Lewis and Clark. Russia had given Astor permission to trade along her coast in North America; the United States had sanctioned the organization of this fur company; so Astor virtually had the backing of two nations in his enterprise. The sea-going party was placed under the command of Lieutenant Jonathan Thorn of the United States Navy, who at that time was on a leave of absence. Thorn was a man of exceptional courage, but lacked the tact necessary to handle such a motley crowd as embarked with him. There were noisy French boatmen, scribbling clerks, and, hardest of all for the doughty Captain to manage, the Scottish partners, who felt that they had the right to give him orders.

After much storm both outside the ship and within it, they arrived at the mouth of the Columbia, March 22, 1811, where in seeking for a practicable passage across the bar nine of the men were lost with a small boat. East of Cape Disappointment the "Tonquin" came to anchor. Two weeks after, at Point George a post was established named "Astoria."[1]

ASTORIA IN 1813

After the cargo was removed the Tonquin coasted to the north in search of furs. Her fate was most spectacular. Not one of the crew returned to Astoria, and what follows was gleaned from the story of the Indian interpreter, Lamazee, who alone of all that left Astoria escaped the catastrophe. It seems that while trading with the Indians near Nootka Sound, Thorn suddenly lost patience with the haggling and bantering of a chief, snatched an otter-skin from his hands, rubbed it in his face, and "dismissed him over the side of the ship with no very complimentary application to

[1] The fort of Lewis and Clark was still standing when Astoria was built, and the names of several of the party were cut upon the logs.

accelerate his exit. He then kicked the peltries to right and left about the deck and broke up the market in the most tempestuous manner."[1] The Indians concealed their rage and in a day or two were back on the ship in overwhelming numbers. They haggled not at all, but took whatever was offered them, being especially eager, however, to trade for knives and hatchets. Having lulled all suspicion to rest, they suddenly, at a signal given, sprang upon the crew, cut down Thorn, and killed all but four men who happened to be in the rigging, and Mr. Lewis, the ship's clerk, who got to the cabin and opened fire upon them. With the help of the four men from the rigging he drove them from the ship. The four sailors tried to escape that night by boat, but were captured, tortured, and killed. Lewis, badly wounded and despairing of succor, resolved to die like Samson in the midst of his enemies. By friendly signs he invited the Indians to come on board, and when they had thronged the deck he lighted the powder magazine, blowing himself, his ship, and his foes to instant death. Lamazee, who had gone ashore with the Indians after the attack, saw all this from the shore, and he seems never to have forgotten the horror of it.

Here was a discouraging situation for the handful of men at Astoria, in a hostile country surrounded by revengeful savages. But McDougal rose to the occasion, and ingeniously contrived a scheme for keeping the Indians quiet. Some years before this smallpox had appeared in this region and almost exterminated entire tribes, so that the Indians had the utmost dread of the disease. McDougal called the chiefs together and showed them a small bottle, saying, "In this bottle I hold smallpox, safely corked up. I have but to draw the cork and let loose the pestilence to sweep man, woman, and child from the face of the earth." The chiefs made haste to assure McDougal that they wanted to be

[1] Irving. Astoria.

friendly and neighborly, and pledged their loyalty to the company, and particularly to the "Great Smallpox Chief!"

While the Astorians were thus trying to establish a foothold on the Pacific coast, the overland party, led by Hunt, was toiling westward under still greater difficulties. With them was Donald McKenzie, one of the partners who had had ten years' experience in the employ of the Northwest Company. The party left St. Louis October 21, 1810, camping for the winter at a spot a little below the site of St. Joseph, Missouri. As early as possible in the spring of 1811 they resumed the journey up the Missouri, hurrying to get out of the way of Manuel Lisa, who had been a thorn in their side while they were near St. Louis. Lisa did not like to see interlopers endeavoring to share with him the profits of the western fur trade. Furthermore, Pierre Dorion, the gifted half-breed interpreter whom Hunt had hired, owed him a whiskey debt, and Lisa felt that until it was paid he should have the first call on Dorion's services. Up the river went Hunt as fast as oar, pole, and cordelle could drag his heavy boats. They passed La Charette, the last squalid village of the whites, and just beyond it found Daniel Boone, still squatting on the farthest frontier. John Colter, too, the man of Lewis and Clark's party who had turned back to the wilderness from the Mandan Villages, was settled near here, and accompanied them for a day, doubtless wishing heartily that he might stay with them to the end; but he had recently taken a wife and could no longer roam at will. With Hunt's party were Nuttall and Bradbury, two scientists going to the unknown land to increase the sum of scientific knowledge. In Bradbury's Journal will you find an interesting account of this meeting with these two heroes, of Colter's thrilling escape from the Blackfeet, and of Boone's skill with the rifle and trap, though he was then an old man of eighty-four years.

Lisa, on his way to relieve Henry, was coming swiftly up

the river, and Hunt, fearing his enmity and eager to pass
through the Sioux and Arikara tribes before Lisa could reach
them and set them against him, strained every nerve to keep
the lead. But by
the second of June
Lisa had overtaken
him, just as they
arrived in the hos-
tile Sioux territory.
Together the two
parties traveled as
far as the Arikara
Villages, which
were below the
Mandan's and a
few miles above the
junction of the
Grand and Mis-
souri rivers. Hunt
had expected to go
to the source of the
Missouri, and then
down the Colum-
bia, following the
trail of Lewis and
Clark; but the ac-
counts of hostility
of the Blackfeet

From Maximilian's Travels
AN ARIKARA INDIAN, PASHTUWA CHTA

decided him to abandon the boats at this point and endeavor
to reach the coast by an overland route. In order to avoid
the dreaded Blackfeet, Hunt and his men turned to the southwest, went through the Dakotas, across the southeast corner
of Montana into Wyoming near the northeastern boundary,
and south into the Wind River country, where they encount-

ered the Shoshone and Flathead Indians.   During the journey the guide and interpreter deserted the party, exposing
the members to unnecessary hardships and to the cunning of
the Crow Indians.   Before leaving what is now Wyoming
they saw the Tetons and named them "Pilot Knobs."   After
leaving Pierre's Hole they found the journey easy enough

THE TETONS AND JACKSON LAKE (Wyoming)

down the Snake River to Henry's post.   At this post Hunt
made the serious mistake of his trip by abandoning his
horses and taking to the river.   Leaving some of his men to
trap and hunt, entrusting his horses to the care of two
Snake Indians, Hunt embarked with the remainder of his
men in fifteen canoes.   The error of this change was soon
evident, for the rapid Snake proved unnavigable.   Abandoning their canoes and hiding their goods in many *caches*, the
forty-nine men on foot pushed toward the west.   The party
divided into three divisions, two following the general direction of the Snake River, while Hunt's path went north to the
present site of Boise City, then down the Columbia to Astoria, February 15, 1812.   Thus we have the first white man
and his party over the Oregon Trail.

Great was the rejoicing at Astoria over the reunion of the sea and land forces. In accordance with arrangements made by Astor, a dispatch was sent to him in New York, telling of the safe arrival of both parties. Robert Stuart was selected to attempt the long journey overland to the east with an escort of only five men. The return trip to the place where Hunt had cached his supplies was much the

THE SNAKE RIVER ISSUING FROM JACKSON LAKE

same as that of the outgoing Astorians. At the place where the goods were cached they met the Snake Indian who had acted as guide from the mouth of Hoback River to Henry's Fort the year before. This Indian also had charge of the abandoned horses. He reported that the horses had been stolen, the caches broken open, the supplies all taken, and the trapping parties left destitute. At Fort Henry Stuart found four of the party who told of their wanderings, and a "doleful narrative it was." From here Stuart and his men pushed on toward the east, going on what afterwards became the Oregon Trail. The one mistake that Stuart then made was to abandon the course he was taking and endeavor to

SHOSHONE FALLS IN SNAKE RIVER, IDAHO, AND TWIN FALLS-JEROME
STAGE ROAD

find Henry's route to the north. This occurred at a point east of Bear Lake on the boundary line between Utah and Wyoming. After going north as far as Pierre's Hole the party journeyed east to Wind River (Wyoming), getting on the Oregon Trail route again, which then was an Indian path.

DEVIL'S GATE BETWEEN INDEPENDENCE ROCK AND SOUTH PASS

Had they followed this path they would have found that great gateway between the east and west, South Pass, through which in the years to come many thousand weary and hopeful travelers were to pass on their way to Oregon, California, and Utah. Missing this, Stuart's party had a heart-breaking experience. Down to the south and west, Stuart went into the barren deserts of the Sweetwater Mountains, then to the Sweetwater River where the party found abundance of buffalo. From here they went by way of the historic

Devil's Gate and Independence Rock to North Platte River. It was now the last of October and winter was rapidly approaching. The remainder of the journey could not possibly be made before the coming of the heavy snows. In view of this fact, Stuart decided to make this point on the Platte his winter quarters. Here his men built a warm cabin in a wooded bottomland opposite the mouth of Poison Spider Creek. "This was the first building within the limits of the present state of Wyoming."[1]

Game was found in great abundance, and the explorers fairly lined their little cabin with the "jerked" meat of the buffalo. Peace was not long to be their lot, however, for the Arapahoe Indians discovered their little hermitage and ate them out of "house and home." The little nook of security being no longer safe, the party pushed down the Platte to where is now Wellsville (Nebraska), several miles down the stream from what was to be Fort Laramie. Here they built another winter hut, occupying themselves with making canoes until early in March, 1813, when they started on the last stage of their long journey. While on their way down the river they received the first news of the war between the United States and England. "In perfect health and fine spirits" they arrived in St. Louis April 30, 1813.

As two years had passed without news since Hunt's party had left St. Louis, the tidings of its safe arrival at Astoria were received with great delight. Hunt made the distance between St. Louis and the coast in three hundred forty days, Stuart in three hundred six days. Chittenden says: "The two Astorian expeditions are entitled to the credit of having practically opened up the Oregon Trail from the Missouri River at the mouth of the Kansas to the mouth of the Columbia River." It is for this reason that it seemed desirable to give so much space to the travels of the Astorian parties.

[1] Chittenden

Upon the outbreak of the war of 1812 the Northwest Company at once took steps to drive the Americans from Astoria. While Hunt was away on a side exploration, leaving McDougal in charge, representatives of the Northwest Company came to Astoria, and received the heartiest welcome from the Scotchmen whose interests had never been entirely alienated from the British Company. When Hunt returned he found Astoria in the hands of the rival company, and the

FORT VANCOUVER AS IT APPEARED IN 1845

fort renamed Fort George, with the British flag flying where the Stars and Stripes had been. McDougal sold $100,000 worth of furs to the Northwest Company for $40,000, and obtained a good position with the rival company. Final settlement was made and the Astoria dream was at an end.

By the terms of the Treaty of Ghent, ending the war with England, it was agreed that all places captured during the conflict should be returned. Thus Astoria was restored to the United States, though England claimed the mouth of the Columbia; and the Hudson's Bay Company, which had absorbed the Northwest Company, built Fort Vancouver opposite the mouth of the Willamette, one hundred miles up the Columbia, and for many years held the mastery of the Northwest.

It was not until 1818 that any definite conclusion was reached, and then it was agreed that all lands north of the 42d parallel east to the Rocky Mountains should be neutral ground and open to both the United States and England. The Spaniards for years had claimed all of the land along the Pacific to the 55th parallel. Russia at the same time demanded all of the territory on the coast down to the fifty-first degree; the United States claimed to the Rio Grande in virtue of the Louisiana Purchase. In 1819 a treaty was made with Spain by which Florida was bought for $19,000,000, and Spain ceded to the United States all her claims to the Pacific coast lying between the 42d parallel and 54° 40', while we gave up all claim to Texas. With this treaty the boundaries of the Louisiana Purchase and the Spanish possessions were definitely settled and located, and all Spanish claims north of 42° became ours. The United States now had a right to claim the Oregon domain through six channels: by the discovery of 1792, when Gray of Boston found the mouth of the Columbia; by the purchase from France in 1803; by exploration through Lewis and Clark in 1805–6; by the establishment of Astoria in 1811; by the journey of Wilson Hunt in 1812; and by treaty with Spain in 1819.

3. THE ROCKY MOUNTAIN FUR COMPANY

In 1821 the Northwest Fur Company was absorbed by the Hudson's Bay Fur Company, and the huge concern proceeded to monopolize the fur business on the west side of the Rocky Mountains. The headquarters were at Fort Vancouver on the northern bank of the Columbia opposite the mouth of the Willamette. At this time General William Ashley of St. Louis, a man of much experience and large business capacity, thought he saw an opportunity to enter into the fur trade on an extended scale. Early in the

spring of 1822 he organized a company under the name of
the Rocky Mountain Fur Company. With Ashley were
associated Andrew Henry, who had trapped for Lisa, Jedediah S. Smith, William Sublette, Milton Sublette, David
E. Jackson, Robert Campbell, James Bridger, Etienne
Provost, and many others, all of whom wrote their names
large in the early history of the West. Many streams,
lakes, mountain peaks, passes, and forts are named after
these brave explorers.

Ashley was a Virginian who had lived at St. Louis since
1802. Having had twenty years of experience on the frontier, he was well qualified to head an enterprise of the magnitude of the Rocky Mountain Fur Company. The first
expedition left St. Louis April 15, 1822, bound for the
Three Forks of the Missouri, the region that Ashley thought
had "a wealth not surpassed by the mines of Peru." After
being out but a short distance one of the keelboats struck a
snag, going to the bottom of the river and taking with it
$10,000 worth of property belonging to the Company. Not-
withstanding this heavy loss, the party went on. Without
further accident they all arrived at the Mandan Villages,
just above which they lost all their horses through a raid of
the Assinniboine Indians. This made it impossible to push
on to Three Forks before the coming of winter, so they built
Ashley-Henry Fort, at the junction of the Yellowstone and
Missouri rivers.

Here they trapped all winter, and in the spring started
for the country of the Blackfeet. These Indians had not
lost any of their fierceness and drove the men back to the
mouth of the Yellowstone. Ashley had returned to St.
Louis, and at this time had reached the Arikara villages on
his way up the river with reinforcements. The Arikaras
were a most unreliable tribe, one day pretending friendship,
the next at war. This time it was war, and after a sharp

fight they killed one of Ashley's men, wounded four others, and obliged the party to retreat down the river, where Ashley called for volunteers to carry a message to Henry. Much to the surprise of every one, Jedediah Smith, the boy of the party, offered his services. With marvelous escapes this stripling reached Henry, who with twenty men promptly came to the rescue of Ashley. From this point the entire

IN THE GROS VENTRE MOUNTAINS NEAR JACKSON'S LAKE

party went down the river to the mouth of White River, where they established Fort Lookout, and waited for the United States troops to escort them beyond the fighting tribe. After receiving assistance from the soldiers, Henry with eighty men, among whom were Smith, Bridger, Provost, Jackson, and Sublette, started for the Yellowstone. At the mouth of the Big Horn River he established another Ashley-Henry Post not far from the site of Fort Manuel. From here he sent out Etienne Provost with a small party to trap to the southwest. It was on this journey, in 1823, that Provost discovered South Pass, an open highway in the central part of Wyoming, the easy road across the Rockies.

In Utah we find a pretty river, a picturesque cañon, and a thriving city, all named after this old partisan of Ashley's, though the name has been shortened to Provo. It was Provost and his party that found the good trapping-grounds in the region of the Great Salt Lake. It was Jim Bridger who found the Great Lake in the winter of 1824. It was Jedediah Smith, in 1827, who first crossed the Sierra Nevada Mountains separating California from the East. Ashley established a post on Utah Lake near the site of the present city of Provo in 1825, and the next year took out a small cannon to be mounted there, the first wheeled vehicle to cross South Pass. The wheels of this engine of war made the first dim traces of the Oregon Trail, that wonderful road that was to lead to the peaceful conquest of the vast region known as the Oregon Country. Great store of beaver were found on Bear River, Green River, Provo River, Weber River, and Utah Lake, and Ashley became a rich man, potent in the politics of Missouri. His bands, led by such partisans as Provost, Bridger, Smith, Jackson, the two Sublettes and Fitzpatrick, penetrated into every nook of this unknown land, trapped on every stream and lake, found every fertile valley and mountain pass. Ashley himself was the first white man to navigate Green River, which at that time was supposed to empty into the Gulf of Mexico, but which came to be known as a branch of the Colorado emptying into the Gulf of California. Down Green River Ashley went, as far as the mouth of the stream now bearing his name. Forty years after this a United States Geological Survey on its entrance into the Red Cañon found inscribed on a high rock, "Ashley 1825."

The most picturesque event in the lives of the fur men was the "rendezvous" held annually in some favored spot, such as Pierre's Hole (now Teton Basin, Idaho), Ogden's Hole, where Ogden, Utah, now stands, or the valley of the Green

River. Every trapper knew where the rendezvous would be held, and about the first of July each year they began to gather. Here would come gaily attired gentlemen from the mountains of the south, with a dash of the Mexican about them, their bridles heavy with silver, their hat brims rakishly pinned up with gold nuggets, and with Kit Carson or Dick Wooton in the lead. In strong contrast would appear Jim Bridger and his band, careless of personal appearance, despising foppery, burnt and seamed by the sun and wind of the western deserts, powdered with fine white alkali dust, fully conscious that clothes mean nothing, and that man to man they could measure up with the best of the mountain men. At this gathering you would find excitable Frenchmen looking for guidance to Provost, the two Sublettes, and Fontenelle; the thoroughbred American, Kentuckian in type, with his long, heavy rifle, his six feet of bone and muscle, and his keen, determined, alert vigilance; the canny Scot, typified by Robert Campbell, who won both health and fortune in the mountains; the jolly Irishman, best represented by Fitzpatrick, the man with the broken hand who knew more about the mountains than any other man except possibly Bridger; and mixed in the motley crowd an alloy of Indians—Snakes, Bannocks, Flatheads, Crows, Utes—come to trade furs for powder, lead, guns, knives, hatchets, fancy cloth, and, most coveted of all, whiskey, that made the meanest redskin feel like the greatest chief.

Fur trading was the prime purpose of these gatherings. Great loads of goods were brought from St. Louis, at first on pack animals, but after Captain Bonneville's time, by wagons; and these were traded to the Indians and the free trappers for furs. The organized bands working for the Rocky Mountain Fur Company received their outfits for the coming year, and their wages for the past year, turned in their catch, and departed again for the beaver haunts. In a few days all were scattered and nothing remained to mark the location of the rendezvous save the charred remains of campfires, well gnawed bones, some empty cans, many empty bottles, and generally a few fresh graves to testify to the maddening potency of the fluid those innocent bottles had held. In 1826 Ashley sold his interest in the Rocky Mountain Fur Company to Jedediah Smith, David Jackson, and William Sublette. The business was in their hands when Milton Sublette, a brother of William, in 1830 took wagons over the eastern end of what became the Oregon Trail. But he did not cross South Pass with them; that distinction belongs to Captain Bonneville.

CAPTAIN BONNEVILLE

Bonneville was a Frenchman in the United States army, who, having heard of Ashley's amazing success in the fur trade, decided to turn trader and trapper himself, obtained a leave of absence from the army, and cast his fortune in the West from 1832 to 1835.

With one hundred ten men and two wagons Bonneville started from Independence May 1, 1832. The journey to the northwest was under military discipline, with Captain Bonneville as the commander-in-chief. The route was the usual one taken at this time, northwest across the plains, up the Platte and Sweetwater, past Independence Rock, Devil's Gate, and through South Pass to Green River. On the west bank of this stream, five miles above the mouth of Horse Creek, Bonneville and Fontenelle built Fort Bonneville, or

"Fort Nonsense," as it was called by the trappers. The Indians compelled Bonneville to abandon this fort and move over to the headwaters of the Salmon River for the winter.

YOSEMITE FALLS, DISCOVERED BY WALKER IN 1833

Indeed, he was almost constantly on the move during the three years and more that he spent in the mountains, and so much of this region did he cover personally or by means of side parties, and so excellent were the maps and reports that he made upon returning to the army that he can justly be reckoned the chief contributor to our store of early geographic knowledge about the Far West. Add to this the fact that his exploits furnished Irving with material for one of the most charming books in our language and you will see that our debt to the worthy captain is large. He journeyed all over southern Idaho, western Wyoming, northern Utah, and even to Fort Vancouver on the Columbia. He sent out many expeditions, the most notable of which, under I. P. Walker, crossed the desert to California, discovered

Yosemite Valley, and explored much of that desolate region lying between Great Salt Lake and the Sierra Nevadas. This expedition of Walker's and the two earlier ones of Jedediah Smith furnished the first trustworthy information about California and how to get there. Bonneville's claim to fame rests upon his consummate address in dealing with his turbulent retainers and his Indian neighbors, and upon his large contributions to geographic knowledge. In the eyes of History his failure as a fur trader is but a trivial matter.

After Jedediah Smith, with his partners, had purchased Ashley's fur business in 1826, he started out on a long and perilous journey toward the Pacific, his wanderings covering a period of three years, into a territory then "wholly unknown to the American traders."[1] Starting from Great Salt Lake, Smith and his fourteen men explored around Utah Lake, into the Sevier valley, down the Colorado, west into Southern California, reaching San Diego late in the fall. Viewed with suspicion by the Spanish, Smith left the country, trapping and exploring in the valleys of the San Joaquin and the Sacramento, and in the spring of 1827 crossed the Sierras and returned to the summer rendezvous near Salt Lake. The next year he again penetrated to California, through trials innumerable. Driven out by the Spanish, he went north into Oregon, but at the Umpqua River his party was attacked by Indians and only Smith and three others escaped. They made their way to Fort Vancouver, where Dr. John McLoughlin, chief factor of the Hudson's Bay Company in those parts, not only relieved their necessities but sent out a party which recaptured their furs from the Indians. The good doctor paid Smith $20,000 for the furs. Following the trails made by Jedediah Smith, one is impressed with the fearlessness and sagacity of this man, little more than a boy, who, almost alone, made such expeditions

[1] Chittenden.

over mountains, across the deserts and up and down the Pacific coast.

Another attempt as unsuccessful as Bonneville's to break into the Rocky Mountain fur trade was made by Nathaniel Wyeth, of Boston, and a company made up largely of New Englanders. They knew little about the mountains and had no conception of the difficulties to be overcome by him who

FORT HALL

would wrest riches from their rocky strongholds. They did not realize what vast stretches of desert had to be crossed, what interminable leagues of cactus and sagebrush they would find, without streams and frequently destitute of game. They knew little of the Indian and his ways, and least of all did they know what hostility they would meet from their white brothers who were already installed in this wild land and felt it all too small to admit any newcomers. We find Wyeth and his Bostonians at the rendezvous in Pierre's Hole in July, 1832. Here there was a fierce fight with the Blackfeet, which gave many of Wyeth's men all they wanted of the wilderness, and fully half of his little force turned back. He went to Fort Vancouver, hoping to establish a trade in furs and fish, but the Hudson's Bay Company would sell him nothing, the ship he expected from Boston never came, and he returned to the East beaten but not yet conquered. In 1834 he came out again with a load of goods for the Rocky Mountain Fur Company. They had contracted for those goods, but just when Wyeth arrived control of the company was passing from Smith, Jackson,

A GROSVENTRE DAGGER

and William Sublette to Fitzpatrick, Bridger, and Milton Sublette. These new partners repudiated the bargain, and Wyeth had a lot of unsalable supplies on his hands. To house them until they could be sold he built Fort Hall on the Snake River, nine miles above the mouth of the Portneuf near the site of the present Pocatello, Idaho. Then he pushed on to Fort Vancouver, but found the Hudson's Bay Company as hostile and powerful as ever, and finally had to give up and go home. He had made a plucky fight against great odds. We remember him for his pluck — a quality worth remembrance in any man — and also for the facts that he built historic old Fort Hall, that he brought the Methodist missionaries, Jason and Daniel Lee, to Oregon, and that after his failure some of his Company took land in Willamette Valley, grouped themselves about the mission, and did the first farming there,— becoming the nucleus for the American settlement that was to wrest all that region from the English, and abundantly avenge Wyeth's wrongs upon the Hudson's Bay Company.

4. THE AMERICAN FUR COMPANY

John Jacob Astor was the American Fur Company. He managed the affairs, put his money into the concern and reaped all of the profits. This company was incorporated in New York as early as 1808, but before this Astor had done extensive trading in the region along the Great Lakes, with his headquarters at Michilimackinac, under the name of the Southwest Company. After operating in this region he organized the Pacific Fur Company with headquarters at Astoria. On account of the war with England in 1813 this trade was ruined on the Pacific coast. In 1816 Astor consolidated all his interests in the American Fur Company, and the western department of this company had its headquarters at St. Louis.

All other fur traders at St. Louis opposed the establishment of Astor's headquarters at this place. In fact he met opposition at every step and on every side, as he was looked upon as a monopolist in the fur trade, and if the truth were told he really had about one half of the fur business in the United States. The sympathy was with the small traders, because the public believed that this large company would ultimately crush whatever lay in its way.

Astor's success is not due to any assistance he may have received from other traders, but to his own sound judgment and cautious business principles. This American Fur Company had one serious rival in the Columbia Fur Company, but this company finally became a part of Astor's company and transacted business under the name of the "Upper Missouri Outfit," or "U. M. O.," and its operations were confined to all of the territory above the mouth of the Big Sioux.

This combination occurred in 1828, and the consolidation of these two companies made the American Fur Company the strongest one in the business. It was at this time that Ashley was reaping his richest harvest in furs, making this newly formed company very anxious to go to the mountains and invade his territory. It was thought best, however, to get into those regions gradually, and establish permanent posts at the strategic points. Kenneth McKenzie built Fort Union at the junction of the Yellowstone and the Missouri, ruling like a king a territory larger than many European kingdoms. This proved to be the best built fort on the Missouri, and, with the exception of Fort Bent on the Arkansas, was the very best in the entire West.

Up to this time the country of the Blackfeet on the Maria's River had been uninvaded. The Missouri Fur Company, Ashley, and Henry had all made unsuccessful attempts, but were always driven back by these fierce warriors. These Blackfeet did all of their trading with the British, who did their best to perpetuate the feud between the Indians and the trappers to the south. But the tributaries of the Missouri in that region were full of rich furs and McKenzie was determined to possess them. McKenzie made a treaty with the Blackfeet through the cleverness and bravery of a noted trapper named Berger, whom he sent with twelve volunteers into the heart of their country. Audacity alone saved their scalps. Berger brought the Blackfeet to Fort Union, though his ingenuity and firmness were taxed to the utmost to keep them from turning back, and he once had to pledge his scalp and all his horses that they would reach the fort in one more day. Reach it they did, and the treaty was made that wrested the larger part of this extensive business from the hands of the British.

The American Fur Company, through detachments sent out at various times from Fort Union, built Fort Piegan at the junction of the Maria's and Missouri; Fort McKenzie, about six miles farther up the Missouri, after the Indians burned Fort Piegan; Fort Case at the mouth of the Big

From an old cut

FORT UNION AT THE JUNCTION OF THE MISSOURI AND THE YELLOWSTONE RIVERS. BUILT IN 1829

Horn, in the Crow country near where Fort Manuel and Fort Henry had stood; Fort F. A. Chardon at the mouth of the Judith, where one of the most woeful of border treacheries was later committed; and, most famous of all, Fort Benton, below the Great Falls of the Missouri, where the thriving

From Maximilian's Travels
THE "YELLOWSTONE," THE FIRST STEAMBOAT TO GO ABOVE
COUNCIL BLUFFS, 1832

town of Benton, Montana, now stands. In other parts they had Fort Laramie near the junction of the North Platte and Laramie rivers, an old post of the Rocky Mountain Fur Company, which became famous in the history of the Oregon Trail; and Fort Pierre, named after Pierre Chouteau of the American Fur Company, where Pierre, South Dakota, now stands; besides nearly one hundred lesser posts in the heart of the fur country.

Keel boats had been used on the Missouri up to 1832, but the enterprising McKenzie believed that a steamboat could be successfully operated on the river, and finally obtained one, which he called the "Yellowstone." In 1832 this boat went as far north as Fort Tecumseh, about three miles above the junction of the Teton and the Missouri. This fort then was named Fort Pierre in honor of Pierre Chouteau, Jr. Never had steamboat gone up the Missouri above Council Bluffs, so 1832 marks the beginning of a new era for the Far West. The next year the "Yellowstone" went as far north as Fort Union. Fort Benton was first reached in 1859. This marks the head of navigation on the Missouri, for the Great Falls are only a short distance above. Thus the steamboat was put into use for the north fur country, and continued in service until the arrival of the railroad.

Mr. Astor wrote from France to Chouteau, one of his managers: "Your voyage in the 'Yellowstone' attracted much attention in Europe, and has been noted in all of the papers here." If the steamboat so impressed the people across the waters, what was the impression on the Indians? Chittenden explains this in the following language: "Its power against current, as if moved by some supernatural agency, excited the keenest astonishment and even aroused a feeling of terror." The steamship to the native was a Fire Boat that walked on the waters; it was alive and must be a creation of an evil spirit.

The fur trade was the most potent factor in the early development of the Far West. The trappers found the paths, and some of them, notably Bonneville, mapped these unknown lands; they tamed the natives; they built the forts; they provided transportation by land and water. In their train were scientists, like Nuttall and Bradbury, and missionaries, like Father De Smet, the Lees, Whitman and Spalding, to civilize and Christianize. When the soldier and the settler came to possess the land they found all these agencies ready made for their use. Not only is the era of furs the most romantic — it is also the most important in early western history.

REFERENCES

Chittenden. The American Fur Trade.
Larpenteur. Forty Years a Fur Trader.
Dellenbaugh. Breaking the Wilderness.
Coutant. History of Wyoming.
Irving. Astoria.
Schafer. History of the Pacific Northwest.
Hough. The Way to the West.
Semple. American History and Its Geographic Conditions.
Irving. The Adventures of Captain Bonneville.
Bradbury's Travels.
Dye. McLoughlin and Old Oregon.

BLACKFEET PARCHMENT BAGS

CHAPTER III

THE GREAT TRAILS

1. THE SANTA FÉ TRAIL

In 1880 the first train over the Atchison, Topeka and Santa Fé railroad reached Santa Fé, and that picturesque road over which the explorers and travelers from the time of Cabeza had journeyed was no longer used for commercial purposes. Thus the old Santa Fé Trail, which had for centuries served the purpose of a highway between the Missouri and the Southwest, had its calling usurped by the iron road.

When we read of the journeys of those indomitable men who hunted in the Northwest for furs, we always find that their effort was to establish friendly relations with Indians, for without this the rewards of their labors were most uncertain. In contrast to this the men on the Santa Fé Trail made every endeavor to avoid the Indians and not to come into direct contact with them.

Five years before Coronado pushed up northeast from Mexico to that will-o'-the-wisp city of Quivira, Alvar Nunez Cabeza de Vaca traveled on what in the years to come was known as the Santa Fé Trail. It was on this march that he encountered the countless herds of buffalo or American bison. In an account of his travels he speaks of the buffalo, saying, "Cattle came as far as this. I think that they are

76

about the size of those of Spain. They have small horns like the cows of Morocco, and the hair very long and flocky, like that of the Merino; some are light brown, others black. The Indians make blankets of the hides of those not full grown. They range over a district of more than four hundred leagues, and in the whole extent of the plain over which they run the people that inhabit near there descend and live on them and scatter a vast many skins throughout the country."

Next we find De Soto in this trail region, as he is supposed to have camped on a spot near where Wichita, Kansas, is now located. This part of the country is also where Coronado pushed on with his picked horsemen to the north for the "city of stones," when most of his men went back to Mexico.

In 1884 some mounds in McPherson County, Kansas, were opened and many interesting relics were found, among them a small piece of steel chain armor. This was the kind of protection that the Spanish soldiers wore during the time of Cabeza de Vaca, and Coronado. "The probability is that it was worn by one of De Soto's unfortunate men, as neither De Vaca nor Coronado experienced any difficulty with the savages of the great plains, because their leaders were humane and treated the Indians kindly, in contrast to De Soto, who was the most inhuman of all of the early Spanish explorers."[1]

It is impossible to state exactly the date when commerce was started between Mexico and the United States by way of the Santa Fé Trail. As early as 1804 La Lande from Illinois carried on traffic with the people to the west of the Missouri. Then there was James Pursley, whom Pike spoke of as "the man of gold nuggets," who also went west to dispose of merchandise. Both of these men liked Santa

[1] Inman. The Old Santa Fé Trail. Crane & Co., Topeka.

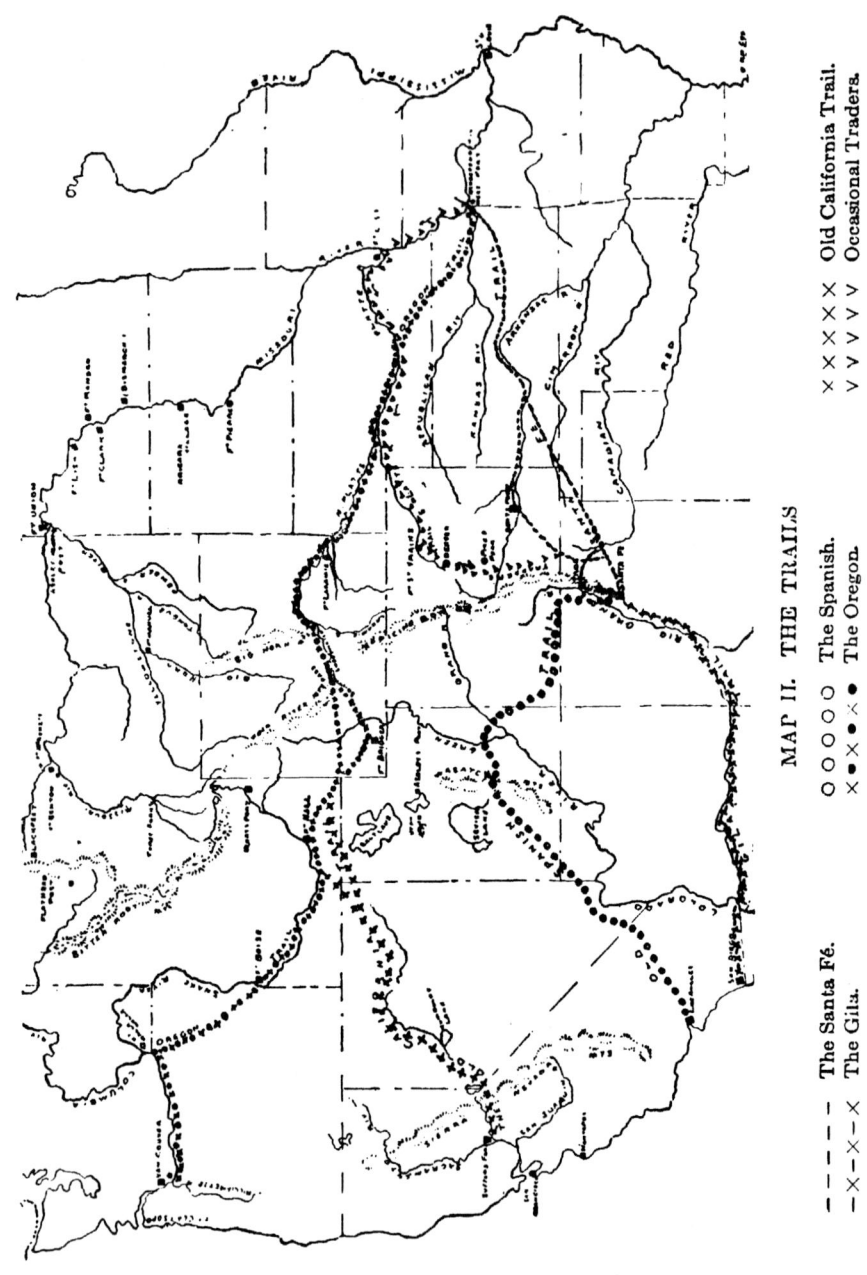

MAP II. THE TRAILS

The Santa Fé.   o o o o o The Spanish.   x x x x x Old California Trail.
x–x–x The Gila.   x • x • x • The Oregon.   v v v v v Occasional Traders.

Fé so well that they never returned to tell of their adventures, but settled down in the land of their adoption.

Pike's expedition taught the people of the United States that a route directly west from St. Louis, by the bend in the Arkansas, was a shorter and better one than by way of the Platte. "It was strong-legged, stout-hearted Zebulon who told of the profits of the possible Spanish trade, and credit is usually given him for first outlining the historic trail along the Arkansas." [1]

In 1812 McKnight, Baird,[2] and Chambers with their associates started for Santa Fé, thinking that the embargo upon trade with the United States had been raised through the Declaration of Hidalgo in 1810. After a weary journey, following the directions laid down by Pike, the party reached Santa Fé only to find that the embargo was in full force. They were seized as spies, imprisoned, and all of their goods confiscated. Here they were kept in strict confinement for over nine years. After their release, in 1821, Chambers and McKnight started at once for the Missouri. On the way home McKnight was killed by the Indians, Chambers making the rest of the journey alone. Baird followed in a few months, making the entire distance without a companion. One would naturally think that after such an experience with the Spaniards one journey into the land of the enemy would be sufficient. Not so with Baird and Chambers, who made a second expedition in 1822. As they started very late in the season they had terrible hardships, and all their animals were frozen to death. The winter was spent near the present site of Dodge City.

Captain Becknell,[3] of Missouri, in 1821 started west to

[1] Hough. The Way to the West. Copyright, 1903. The Bobbs–Merrill Co.

[2] By some authorities, spelled Beard.

[3] Also spelled Bicknell.

trade with the Indians, but when on the headwaters of the Arkansas was prevailed upon by a party of Mexicans that he had met in the mountains to take his merchandise to Santa Fé. The exorbitant prices obtained for goods at Santa Fé started commerce to the West. To illustrate what prices were at that time it is only necessary to state that calico, and this of the most common kind, brought $2 and $3 a yard. Becknell returned to Santa Fé the next year with $5,000 worth of goods of all descriptions.

After arriving on that point of the Arkansas now called the "Caches," Becknell started out southwest over the Cimarron desert, a much shorter though a more dangerous route. From this point of the journey he must be considered a pathbreaker, and he has for this reason been called the "Father of the Santa Fé Trail." That is to say, Becknell first made the trail across the Cimarron desert, although others had broken the way from the Missouri to the "Caches." The Caches derived its name from the fact that Becknell "cached" some of his goods at this point when he had too many to make the desert trip.

Becknell's party suffered intensely from thirst, for the small supply of water carried in their canteens soon became exhausted. So great was their need that they killed their dogs in order to drink the blood, and even cut off the ears of their mules for this purpose. They also went so far as to kill a buffalo that wandered across their path, and drank the water from its stomach, which they said "was an exquisite delight." Going in the direction from which the buffalo came, they found a river, quenched their thirst, and returned with filled canteens to those who were too weak to follow them to the river.

Finally the party reached Santa Fé, being the first white men to make the journey through this terrible desert, also the first to make the trip in wagons. Others followed in this path, and in spite of all their care to provide water many parties endured great hardship and even death on the desolate Cimarron Trail.

Jedediah Smith, who had made such perilous trips to San Diego and Vancouver, lost his life in this desert while hunting for water. Smith, Jackson, and Sublette sold their interest in the Rocky Mountain Fur Company, went south to engage in the Santa Fé trade, and on their very first venture were lost in the Cimarron desert. The party was dying of thirst when Smith struck out alone to find the needed water. After hours of toil in the burning sun he crossed the bed of the Cimarron River, but could not find a drop of water. Still Smith knew the desert streams, and kneeling down scooped up a handful of sand and made a hole in the bed of the river. At once the water commenced to seep into the hole, and there was soon enough for him to satisfy his burning thirst. Just as he was leaning over to drink, a band of Comanches came from their hiding-place and filled his body with arrows.

The date of the earliest settlement of Santa Fé is unknown. Historians do not agree upon the exact time of its founding. Some claim that Cortez founded the city, others that to Coronado belongs the honor. Of this one thing we may be sure, that before Jamestown was settled, or before the Pilgrims made their landing on the Atlantic coast, there existed a town in the Southwest known as Cuidad Real de San Francisco de la Santa Fé. The date of the founding of the city is variously placed at 1540 to 1616. A careful statement made by one who has extensively studied the question places the date at 1605.[1]

There is but little recorded of the early history of this ancient dwelling-place. The Spaniards doubtless found the aborigines easy to conquer, and ready to submit to the

---

[1] Dellenbaugh. Breaking the Wilderness. G. P. Putnam's Sons.

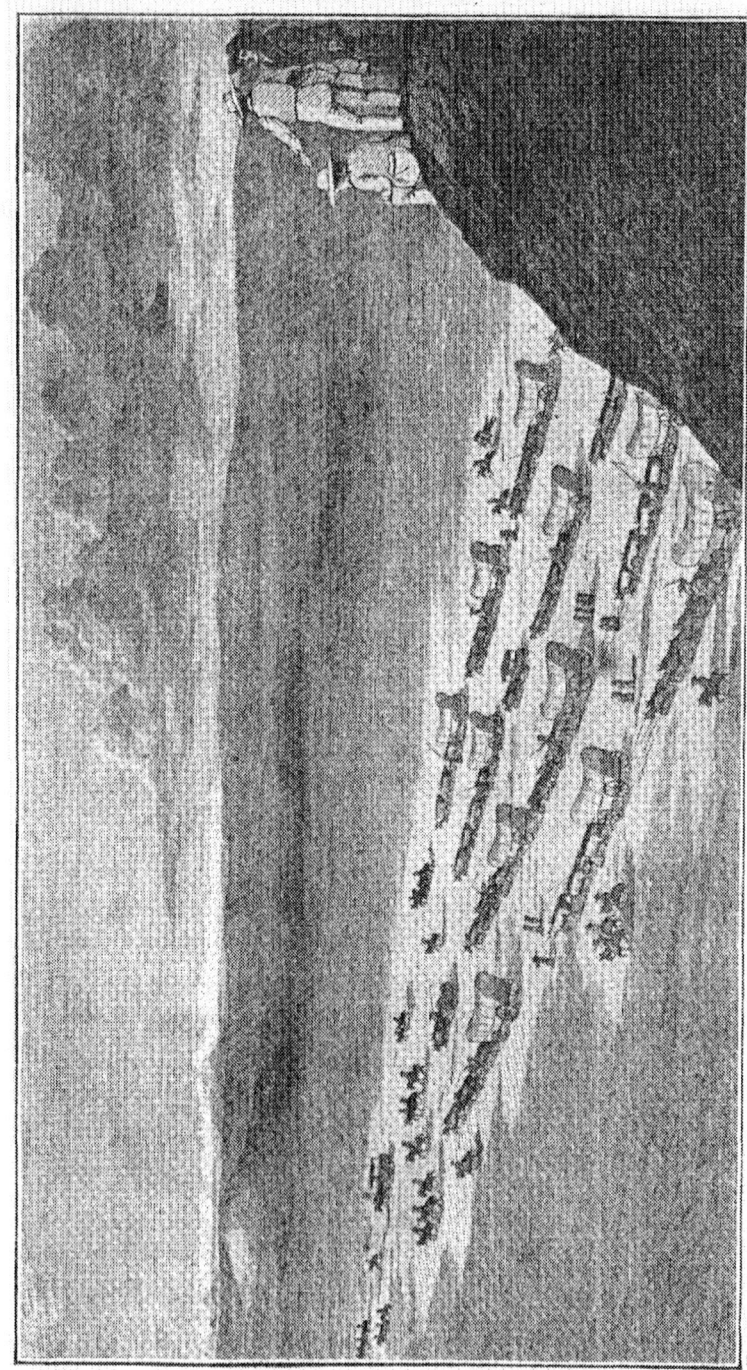

MARCH OF THE SANTA FÉ CARAVAN

authority of those who had subdued them. Conditions continued peaceful for many years, but the heavy tasks imposed upon the Indians, who worked in the mines, aroused in the hearts of these natives a spirit of revolt, and with this came the desire to possess their lands again. Some of these Indians believed that they had the blood of Montezuma in their veins, and could by fighting regain their homes and their liberty. In 1680 a terrible insurrection took place not only at Santa Fé but generally in that part of Mexico. All of the churches of Santa Fé were sacked, vestments stolen, altars destroyed, and monasteries burned. The Spaniards were then driven from New Mexico, and for twelve years the Indians remained in possession of their lands.

De Vargas at the head of a small force reconquered the territory, and henceforth a more humane treatment of the Indians resulted in better conditions generally. The Indians, however, never seemed thoroughly subdued. They were allowed their own government and certain tracts of land, but for many years were restless and hated their conquerors. In 1837 they united with the Mexican insurgents in another bloody battle against the Spaniards. Much of the severity with which the American traders were treated is explained by the fact that the Spaniards believed that they incited rebellion among the Mexicans and Indians.

After the Santa Fé Trail was established there was constant danger for the individual trader making the journey alone. As a result the caravan system was adopted. The caravan at first consisted of a train of loaded mules and burros, pack-animals as they were called. This limited the traffic, as only a comparatively small load could be taken by each animal. With the coming of the wheeled vehicles in 1824 a new impetus was given to the trade over this trail. Now regularly organized companies carried on the traffic

Gregg, *Commerce of the Prairies*

ARRIVAL OF THE CARAVAN AT SANTA FÉ

with great wagons drawn by oxen, mules, and horses, and hauling $30,000 worth of merchandise in a single trip. By 1843 the trade had reached its height, and thereafter declined rapidly, but during the twenty years of its existence it furnished some of the most striking episodes in the early history of the West.

As St. Louis was the headquarters for outfitting the early explorers and the fur traders, so Independence[1] was the starting-place for the trade of the Santa Fé and the Oregon trails. Here goods sent from St. Louis by boat were transferred to wagons or pack animals. Soon Westport, now Kansas City, became the point of transshipment, because it had a safer boat land-ing. This bend of the Mis-

MARKER ON SANTA FÉ TRAIL

souri fixed the location for a large city, and here is one of the best examples of the influence of geographic facts in determining city growth.

For forty miles both of the great trails ran over the same road. Then travelers came to a post with the sign, "Road to Oregon." Thus simply was announced a road over 2,000 miles long. To the right led a road that became a transcontinental highway for the early settler; to the left ran a transcontinental trade-route,—either leading to perils and hardships; both to possible wealth and power.

[1] Situated five miles east of present Kansas City.

2. THE GILA ROUTE AND THE OLD SPANISH TRAIL

Not all who went to Santa Fé with their merchandise over the Trail returned to the States. Many of the merchants or their agents settled in that ancient city, where they carried on further commerce. This trade was carried on chiefly to the south and west. To the south the wagon trains went as far as Chihuahua. The trade at this place became particularly active and profitable, so that between 1830 and 1840 it absorbed nearly one half of the Missouri caravans. To the west from Santa Fé there were no market places such as had been found to the south; but the trapper, following the streams, found the beaver in abundance. Crossing the Continental Divide from the valleys of the Rio Grande and the Arkansas, he came to the Gila, which he could follow without great difficulty to the Gulf of California, or to the Gunnison, which emptied into the Grand, which latter conducted him to its junction with the Green. From this point the united stream is known as the Colorado, and becomes useless as a highway because of its mad race through its wonderful cañon. Here he struck boldly out across the desert, crossed the Sevier and the Virgin rivers, skirted Death Valley, and so through weary leagues of parching desert came to the smiling California valleys, and finally Los Angeles. This was known as the Old Spanish Trail. The one along the Gila was the Gila Trail. Both aimed for southern California, where horses and mules were cheap and where cloth and metal goods brought a good price. In early days the famous copper mines of Santa Rita, along the Gila, had attracted the Spanish, Americans, and Mexicans, who with their pack-horses made well-beaten paths toward the waters of the Pacific. Thus the miner and the trapper encountered one another along the same stream. It was not a difficult or a dangerous undertaking to push

on out to the coast, and thus ultimately the Gila Route came
to be the pathway to San Diego. The shortest way to California from St. Louis was over this route by way of Santa Fé.
It was along this line that the United States Government
in 1846–47 made surveys for a transcontinental railroad.
Sylvester and James Ohio Pattie, father and son, trapped
along the Gila, journeyed to San Diego, and by this means
established regular commercial relations between St. Louis,
Santa Fé, and California. The Gila Trail after this was
quite extensively used, for it proved to be comparatively
safe from the attacks of the Indians. Kit Carson took this
route for the States in 1846 when he carried Fremont's
message for the authorities at Washington. Over this same
trail General Kearny hurried to California to assume military control of the Pacific coast.

· Thus we have established a southern transcontinental
route leading people to the southern coast of the Pacific.
While these trading expeditions were going over the Santa
Fé, Gila, and Old Spanish trails, Oregon Territory was attracting real settlers over that long and dangerous route
known as the Oregon Trail.

3. THE OREGON TRAIL

To Wilson Price Hunt and his expedition for the Pacific
Fur Company must be given the distinction of being the
first explorers over this famous route. Though Lewis and
Clark had twice gone over the country lying between the
Missouri and the Pacific, they were never near or upon this
trail until they came to the Columbia River west of its
junction with the Snake.

For a distance of forty miles from Independence the Santa
Fé and the Oregon trails were one and the same. After that
their paths became farther and farther apart, until one with

its western extension reached the southwest portion of the United States, San Diego and Los Angeles; while the other pushed up to and beyond the Rockies until it reached the other extremity of our possessions on the Pacific coast.

The Oregon Trail followed the route of least resistance, for it was the path of wild animals. Here was first found the narrow and well-beaten path made by the first possessors of the country, the buffalo, the antelope, the elk, and the deer; in their path came the Indian, who was followed by the trapper, who in his turn had the explorer at his heels, to be followed by the pioneer, the settler, the wagon road, and at last the railroad. This is the history of the building of the most of the Oregon Trail—beast, Indian, pack-train, wagon, locomotive. So when Hunt, the first of white men on this trail, made his footprints toward the west, he found those of the Indian pointing the same way, and the Indian had only followed the tracks made by the buffalo and the bear.

The next pathmaker on this trail after the Astorians had done their work was Ashley, who came in 1823 with his Missouri men and trappers, and it was one of Ashley's men, Etienne Provost, who discovered South Pass, the most significant find in the history of the trail. Ashley was followed by Bonneville with his wagons in 1832, then in 1833 came Wyeth, who built Fort Hall, the first resting-place along the road. Robert Campbell and William Sublette in this same year built Fort Laramie, another supply station and place of safety, and in the years to come the most famous resting-place along the route.

With the early trappers on this trail was James Bridger, one of the party that discovered South Pass, and he it was in 1843 who built the first post on the Oregon Trail intended from the first for the use of emigrants, Fort Bridger in the southwestern part of the present Wyoming. Laramie,

FORT BRIDGER

Bridger, Hall, Boise — these were the forts that served as stations on the Oregon Trail, and they were hundreds of miles apart.

Over the trail, which by this time was getting so well marked as to be known to the Indians as the "Great Medicine Road of the Whites," went Whitman and Spalding with their brides, the first white women to traverse this wild land. Father De Smet came in 1840, followed by Fremont in 1842 and Parkman in 1846. By this time people were coming by

FORT BOISE

the hundreds and then they came by the thousands for these were the days of the Mormons, who in 1847 sought the "land of promise" in the territory of the West, by way of the Oregon Trail. After them came the "forty-niners" in their mad rush for California, the land of gold. Even though all of these travelers did not journey over the trail

CHIMNEY ROCK IN 1910. EZRA MEEKER RETRACING THE OREGON TRAIL (Nebraska)

for the entire distance, they helped to make its path deeper and more lasting, until it became so broad and deep that all the years since that time have failed to erase it.

Francis Parkman, fresh from college and wishing to obtain some new material for his literary work, went over the Oregon Trail in 1846, spending a part of this summer at Fort Laramie and the rest of the time in one of the villages of the Sioux, learning their manner of living, their language and customs. This first-hand information was embodied in his "Oregon Trail." No account of this trail can be complete without reference to this classic piece of literature.

When Parkman went over the trail it was not new, for there were numerous westbound wagons ahead of him and as many back of him, all going in the same direction. Near Fort Laramie he met one of Daniel Boone's grandsons, and traded horses with another grandson. These Boones were well scattered along the trail, for we find one of them in Oregon, Chloe Boone, who married one of the Governors of

FORT LARAMIE

that state. One camped on the present site of Denver and negotiated the sale of Colorado for the Indians to the United States; one was in California, and one in Texas.[1]

In the years when this host of emigrants went over the trail it became littered with bedding, stoves, furniture, trunks, all thrown away by the travelers to lighten their load. Parkman says: "It is worth noting that on the Platte one may sometimes see the scattered wrecks of ancient clawfooted tables, well waxed and rubbed; or massive bureaus of carved wood. These, some of them no doubt the relics of ancestral prosperity in colonial times, must

[1] Dye. The Conquest.

THE OLD AND THE NEW   EZRA MEEKER, IN 1910, GOING OVER THE OREGON TRAIL IN THE SAME STYLE
OF WAGON AND WITH OXEN AS HE DID IN 1850   ON THE PLAINS IN WYOMING

have encountered strange vicissitudes. Brought, perhaps, originally from England; then, with the declining fortunes of their owners, borne across the Alleghanies to the wilderness of Ohio and Kentucky; then to Illinois or Missouri; and now at last fondly stowed away for the interminable journey to Oregon. But the stern privations of the way are little anticipated. The cherished relic is thrown out to scorch and crack on the hot prairie." [1]

There are a few well-known landmarks on this trail that deserve special mention. On the Platte just east of the boundary line between Nebraska and Wyoming is the prominent geological formation called Chimney Rock, with cylindrical rocks piled up like a tower. This can be seen for many miles. About one hundred miles west of this is historic Fort Laramie. This fort, the Laramie Plains, Laramie Mountains, and Laramie River all derive their name from the French Canadian trapper Jacques La Ramie, often called Joseph Laramie. As fate would have it, this trapper lost his life about 1820 near the mouth of the river that bears his name.

This fort at different times has borne the name of Williams, John, and Laramie. When it was established it was the center of trading with the Ogalalla bands of the Sioux, and with the Cheyennes and Arapahoes. After purchasing the post the American Fur Company expended about $10,000 for improving it, making it larger and better able to withstand the attacks of the natives. The overland travelers, particularly at the time of the gold excitement, were often attacked by the Indians, and for this reason the Government turned the post into a military fort. This fort was built of sun-dried bricks with walls twenty feet high and four feet thick, enclosing a space two hundred and fifty feet long and two hundred feet wide. Within this enclosure were a dozen

[1] Parkman. The Oregon Trail. Little, Brown & Co.

or more buildings, including a blacksmith shop and carpenter shop, meat and ice houses, and a corral large enough to accommodate two hundred horses.

On the trail just west of the fort there is another striking landmark. This is Independence Rock, which is situated eight hundred thirty-eight miles from the town of Independ-

INDEPENDENCE ROCK, THE "DESERT REGISTER," AND THE
OREGON TRAIL TO THE EXTREME RIGHT

ence, at the commencement of the trail. This rock marks the beginning of the Sweetwater valley. This name was given to the stream because its water tasted sweet compared with the alkali waters that abounded in the region. Independence Rock is visible for many miles before it is reached. It covers about twenty-seven acres, and on its sides to-day may be seen hundreds of names placed there by the travelers on the trail. Father De Smet called the rock "The Great Register of the Desert." About one hundred miles west of this rock, and one thousand from the commencement of the trail, is South Pass, practically marking

the central point between the Missouri River and the Pacific
coast. This is a pass in the Continental Divide, not a narrow
opening, but a broad valley through which wagons can pass
with perfect ease. When South Pass is reached there is
again "a parting of the ways" for at this point the waters
flow toward the west and toward the east.

After leaving Fort Laramie, there was no stopping-place
for supplies until Fort Bridger was reached. Here Bridger
made all kinds of black-
smith repairs, and sold
supplies to the emi-
grants, and many were
needed after a thousand
miles of travel. This post
was also used for trad-
ing with the Indians.
After this the next
stopping-place was Fort
Hall, nine miles above
the junction of the Por-
tneuf and the Snake,
the first station on the
waters of the Columbia.
Many wagons going

FIRST STONE ERECTED IN NEBRASKA
TO MARK THE OVERLAND TRAIL
1811, Stuart of the Astoria Company.
1869, Completion of the Union Pacific Railroad.

over the trail were left at this fort, where pack-horses
were substituted. The Hudson's Bay post, Fort Boise, near
the mouth of the Boise River, always meant to the explorer
that there were only five hundred miles left to travel
before reaching the journey's end at Fort Vancouver.

The Santa Fé Trail was seven hundred seventy-five miles
long; the Oregon Trail two thousand twenty. The Santa
Fé Trail remained a trade route to the end; the Oregon
Trail, almost from the first, was a colonist's route. The
Santa Fé Trail, proper, had little to do with mountains; the

Oregon Trail crossed three great ranges. The Santa Fé Trail was harassed by three tribes of Indians; the Oregon Trail by ten. The Oregon Trail was very much the longer and more difficult, but it was proportionately more useful in the development of the Far West.

OLDEST CUSTOM-HOUSE IN CALIFORNIA, MONTEREY
Over this building the flags of three nations floated, Spain, Mexico, and United States.

4. THE GREAT SALT LAKE AND CALIFORNIA TRAILS

The Old Salt Lake Trail is practically over the same road as the Oregon Trail. Over this trail the Mormons passed, save that they left the Missouri at Council Bluffs instead of Independence, made their own road to Fort Laramie, and after leaving Fort Bridger struck southwest and came into the valley of Great Salt Lake through Emigration Cañon.

The California Trail began on the Oregon Trail near the northern spur of the Wasatch Mountains, on the bend of the Bear River, several miles east of Fort Hall. From here the trail took a southwest turn and passed the north end of the

Great Salt Lake, followed the north side of Humboldt Lake, through the Sierras to the junction of the American Fork and the Sacramento rivers. At this point the trail crossed Smith's path of 1828 when he went to Fort Vancouver from Monterey. This trail also followed Walker's path to a point near Carson. After Salt Lake City became a city of importance all parties bound for California over the Oregon Trail went there to lay in supplies to last them for the final eight hundred miles. Thence they went south of the Great Lake through Rush Valley, and westward across the desert to the Sierras. Here they fell into Walker's or the Old California Trail, and so crossed to Sacramento. Those who wished to reach southern California kept south from Salt Lake City by Utah Lake, and then moved via the Old Spanish Trail to Los Angeles and San Diego. The route by Rush Valley and across the desert became well traveled, and soon had stopping-places at convenient intervals, for over it went the overland stage and the pony express.

REFERENCES

Prince's Historical Sketches of New Mexico.
Dellenbaugh. Breaking the Wilderness.
Inman. The Old Santa Fé Trail.
Gregg. Commerce of the Prairies.
Hough. The Way to the West.
Bancroft. History of California.
Thwaites. Edition of Pattie's Personal Narrative.
Semple. American History and Its Geographic Conditions.
Chittenden. American Fur Trade.
Parkman. The Oregon Trail.
Inman. The Great Salt Lake Trail.
Irving. The Adventures of Captain Bonneville.
Paxson. The Last American Frontier.

CHAPTER IV

THE MISSIONS

1. THE CATHOLICS IN THE SOUTHWEST

The Catholics more than any other church or sect have conquered by Sword and Cross. They have subdued and Christianized at the same time. The religion of the Spanish Catholic kept pace with his battles, hence no history can be written of those early settlers in the Southwest without giving attention to the missionary pioneers.

It has been said that the reason that the Pueblo Indians enjoy to-day the possession of their lands is that the Spaniards made special laws for the natives by which they might have undisturbed possession of their lands for all time. The natives after being brought under the rule of Spain were taught obedience, and with that obedience came protection to them in their homes and for their families. To this the Spaniards added a process of educating and Christianizing, although it was a difficult task to make the natives reject their old form of religion and adopt that of the alien race.

"The religions of our North American Indians had many astonishing and dreadful features; but they were mild and civilized compared with the hideous rites of Mexico and the southern lands."[1] Their religion was one of fear, revenge, human sacrifice, and idols. It was this form of religion that

[1] Lummis. The Spanish Pioneers. A. C. McClurg & Co.

98

the missionaries had to contend with and attempt to sub-
stitute for it their religion of love, obedience, and brotherhood.

Fray Marcos, who first went to the "Seven Cities" scouting for Coronado in 1539, was a true type of the pioneer
missionary. The field in which he did his missionary work
was what is now New Mexico and Arizona. It was a lonely
life, full of sacrifice and incredible hardships, that this
apostle led in the desert. Indeed, all of the priests who were
brought to New Mexico by Coronado did true missionary
work; they had no hope for worldly reward; their compen-
sation lay in the consciousness of God's work well done.

The Southwest has many old ruins of churches which
were built over three hundred years ago. As far back as
1598 there was a small chapel, the second church in the
United States and the first in what is now New Mexico, built
by the missionaries at San Gabriel de los Españolas, where
Chamita is now located. It must be remembered that three
years before the Mayflower came to America Christian service was being held in at least eleven churches in the region
now called New Mexico, and in one other church a hundred
miles toward the Pacific.

Those who sought to make homes in the desert of the
Southwest had to confine their operations to both sides of
the Rio Grande in order to obtain sufficient water, but the
missionaries knew no boundary lines and pushed out into
the desert in all directions. As early as 1629 these fearless
and God-fearing men had penetrated the wilderness to Zuni,
many miles west of Santa Fé. To this day one of the
churches built by them is in splendid state of preservation.

On the present boundary line between Mexico and the
United States Fray Garcia de San Francisco, in 1662, built
a church long, lonely miles from any Spanish settlement.
Not only did the Spaniards build along the boundary lines;
but half a century before our nation was born they built

in one of our territories, New Mexico, "half a hundred permanent churches, nearly all of stone, and nearly all for the express benefit of the Indians. That is a missionary record which has never been equalled elsewhere in the United States."[1]

These priests all came from Spain by the way of Old Mexico. After reporting to the higher authorities at Santa Fé, they were assigned a mission which might be near at hand or might be many hundred miles away. To a lonely spot the missionary priests made their journey, generally unescorted, across the trackless desert. At a village of savage natives the priest would establish his little chapel, and attempt to teach the gospel in a tongue totally unknown to his listeners, whose language was equally unknown to him.

In this lonely place, far removed from all trace of civilization, the priest was absolutely at the mercy of these strange, foreign-speaking people. If they wished to starve him, or even to kill him, he could only submit. No word would ever get back to his church how he had died defending the Faith. This is the highest type of sacrifice, that one die for principle and no one know it.

It has been from these "Fathers" that we have obtained some of our most valuable early history of the southwest country. These priests who were sent out to the frontier were not of an ignorant class, but were educated and able to observe and record what went on about them. They wrote of the way the natives lived and of the language they spoke.

Within the present boundaries of California, the first Catholic mission was established at San Diego in 1769. This was during the time when the colonies on the Atlantic coast were remonstrating with England over her tax on tea, glass, paints, and paper. Little these pious people thought

[1] Lummis.

or knew of the religious work that was going on at the other side of the continent. Among the Spanish "Padres" was Padre Junípero Serra, a member of the Franciscan order, who established this first mission at San Diego in July, 1769, and during the same year also founded the San Carlos

SAN CARLOS MISSION, FOUNDED BY PADRE JUNÍPERO SERRA, 1769, MONTEREY

Mission at Monterey, where he spent his last days in beginning to colonize California, teaching the Indians the Word ,of God and the rudiments of agriculture. When Father Serra died in 1784 he was at the head of all of the missions in California, and noted for his organizing ability and saintliness.

All of the missions established by the Catholics north of San Diego were known as the Northern Missions. To these missions the name of California ultimately was attached,

a name which had long been given to those on the peninsula
south of San Diego, while Upper California was called Alta
California.   The Jesuit order in Lower California was superseded in 1767 by the Franciscan order.   This order at once
attempted the establishment of many other missions, and
founded San Diego in 1769.   To the harbor of San Diego

Southern Pacific Railway
SANTA BARBARA MISSION, FOUNDED BY PADRE JUNÍPERO
SERRA, 1786

these priests took their domestic animals and formed a
permanent settlement.   From here they went north, and
in time established missions at Santa Barbara, Monterey,
San José, Santa Cruz, and San Juan.   In fact, before the end
of the century the Catholics had established missions all
the way between San Diego and San Francisco.

There was no overland communication between the
missions of New Mexico and California, although attempts
were made to establish connection by the way of the desert
and mountains.   As the missions of California had grown

rich and self-sustaining, there was no special need of communication with the world directly east of them. When communication was established some sixty years later it was the beginning of the end. The long years of Spanish possession in the Southwest had served to prepare the way for another people, more pushing and aggressive, who were destined to found here one of the richest states of our nation.

2. THE METHODISTS IN OREGON

There were notable early laborers in the country drained by the Columbia. They went there originally at the request of the Indians, who wanted the "white man's religion" brought to their tribes. It has been said that the missionary work with the Oregon Indians began in a romance and ended in a tragedy. Possibly the following story of the Flathead braves is no more than a romance, but there is good reason to believe that something of the sort happened.

For many years after Lewis and Clark had passed through the country of the Flatheads and the Nez Perces, the Indians talked of the white man and his ways. In some way an Indian had obtained possession of a high silk hat at the time Lewis and Clark's expedition went west; and for thirty years this old, high, black headgear was a symbol of the white man. The hat was passed from head to head as a badge of honor, and he who wore it was very proud. The Indians said that Lewis and Clark also carried with them a long, straight piece of iron from which they could command the thunder and lightning, and that, in addition to this awe-inspiring rod, the explorers carried a "brass voice" that could make a noise louder than the howl of the bear or the roar of the buffalo. These mysterious instruments were to the Indians a sign that the white man was the favorite child of the "Great Spirit." They also said that the white man had a "Book of Heaven" which told him how to live happily, and how to reach the happy hunting-ground after death.

It seems that the Flatheads finally decided to send for that wonderful book, for in the fall of 1831 four Flathead braves appeared in St. Louis, saying that they had journeyed on foot for many moons to reach the white man to beg for the book, and for teachers who might show them how to use it. Captain Clark, the governor, who was then in charge of Indian affairs for all the region west of the Missouri, sent them to the various churches of the town. They were fêted and made much of, taken to theaters, balls, receptions, and great dinners, until one of the simple sons of the mountains, Black Eagle, succumbed that fall to the strain and was buried at the Catholic cathedral in St. Louis, far from his home and his people. In the early spring another of the chiefs, The Man-of-the-Morning, died, leaving the two younger men to convey the message from civilization to their tribes. These two chiefs started home in the spring of 1832, on the steamboat "Yellowstone," when it made its first trip up the Missouri, but only one, Rabbit-Skin-Leggins, lived to reach his people, No-Horns-on-His-Head having died on· the journey somewhere near the mouth of the Yellowstone.

Let us hope that the sole survivor was the chief that made the eloquent appeal which stirred the religious people of the East into action. This fine specimen of Indian eloquence was delivered at a farewell dinner given by Captain Clark, and it galvanized into life the sleeping missionary spirit of the East. With a mournful sense of failure, the chief said: "I came to you over the trail of many moons from the setting sun. You were the friends of my fathers, who all have gone the long way. I came with an eye partly open for my people, who sit in darkness. How can I go back blind to my blind people? I made my way to you with strong arms, through many enemies and strange lands, that I might carry back much to them. I go back with both arms broken and empty. Two fathers came with us; they were the braves of many winters and wars. We leave them asleep by your great water and wigwams. They were tired in many moons. My people sent me to get the white man's book of Heaven. You make my feet heavy with gifts and my moccasins will grow old in carrying them. When I tell my blind people, after one more snow, in the big council that I did not bring the book, no word will be spoken by the old men or the young braves. One by one they will rise up and go out in silence. My people will die in darkness, and they will go a long path to other

DR. JOHN McLOUGHLIN

hunting-grounds. No white man will go with them, and no white man's book to make the way plain."

The Methodists were first to respond to this stirring appeal in 1834. They sent Jason and Daniel Lee[1] to open a mission, in "Oregon," which then embraced a vast region since divided into the states of Oregon, Washington, Idaho,

[1] Associated with the Lees were Cyrus Shepard, C. M. Walker, and P. L. Edwards, who journeyed west with the little band from the Missouri River.

western Montana, and northwestern Wyoming. Wyeth
was just starting for the West on his second trip at this time,
and the missionaries accompanied him. The Lees did not
hasten directly to the Flatheads upon reaching the land of
the Columbia, but decided, after consultation with that
sage adviser, Dr. McLoughlin, to settle on the fertile coast
lands, where greater protection could be afforded them and
where there was at least as much need of their services as
among the Indians in the more desolate mountain regions.
In the beautiful valley of the Willamette were many Indians,
and French Canadians with Indian wives and half-breed
children, who needed not only the "White Man's Book"
but also schools and other civilizing forces.

So these spiritual pioneers journeyed beyond the land
of the Flatheads and settled in the Willamette Valley, not
far south of Fort Vancouver. Their mission became a
nucleus for American settlement, and their Indian school
developed into Willamette University. This was the
humble beginning of American occupation that finally
thwarted the plans of the Hudson's Bay Company and
saved Oregon to the Union.

3. WHITMAN AND SPALDING

The American Board of Commissioners for Foregin
Missions, representing both the Presbyterians and the
Congregationalists, was next to act. It sent out in 1835
Rev. Samuel Parker and Dr. Marcus Whitman to look over
the Oregon Country and report as to the advisability of
sending missionaries to that far-off land. As the Lees had
gone out in the train of Wyeth on his second expedition, so
Parker and Whitman attached themselves to a party of
the Rocky Mountain Fur Traders, led by Fontenelle.
When they reached South Pass in August, and learned that

they were really crossing the Rocky Mountains, Parker was prompted to write: "There would be no difficulty in constructing a railroad from the Atlantic to the Pacific."[1] Doubtless, the good old man found reason to revise his statement when he got farther west among the precipitous slopes of the Blue Mountains and the chasms of the Snake River. This idea of a transcontinental line was first suggested by Robert Mills in 1819, thirteen years after Lewis and Clark had returned from their expedition.[2]

The missionaries preached to a motley assemblage of mountain men and Indians at the Green River rendezvous in the summer of 1835. So sincere seemed the interest of the Indians assembled here, and so evident was their need of help and guidance that it was resolved that one of the missionaries should return from this point and obtain the help necessary to found a mission, while the other went on and determined upon a desirable location. The old man, Parker, went on into the wilderness. The young man, Whitman, returned to civilization, but not until he had vindicated his right to the title of doctor by removing from the back of Jim Bridger an arrowhead that had been embedded in his flesh for three years. Taking two young Nez Perce Indians as "specimens" the Doctor went back with the wagons that had come to the rendezvous with supplies from St. Louis and were now returning laden with fur; while Parker, with the able guidance of Jim Bridger himself, went on to the land of the Nez Perces, rambled about over their beautiful hunting-grounds, visited forts Walla Walla and Vancouver, and finally took ship for Boston by way of Honolulu.

Dr. Whitman returned in the spring of 1836, accompanied by his bride and by Mr. and Mrs. H. H. Spalding, another

[1] Of all passes through the Continental Divide, South Pass is the only one through which no railroad yet runs.

[2] Wheeler. The Trail of Lewis and Clark. G. P. Putnam's Sons.

recently married couple. A strange bridal tour it must have
been for these two young women, the most remarkable one on
record. With a party of fur traders as escort, they journeyed
in rude wagons, the first wheeled vehicles to traverse the entire
Oregon Trail, across the interminable leagues of sun-parched
plains, through tribe after tribe of savage redmen, who
crowded about in awe to see these wonderfully fair creatures

FORT WALLA WALLA

the first white women they had ever seen; forded the Platte,
the Sweetwater, the Green, and many lesser streams;
scrambled through mountain passes; and finally settled
down to their life work amid the rudest surroundings, the
Whitmans at Waiilatpu, not far from the present town of
Walla Walla, the Spaldings about one hundred twenty
miles up the Clearwater, east of the present town of Lewiston,
Idaho, at a place called Lapwai. With rude log buildings
hastily constructed to shelter them and their stock, the
two families settled down to instruct and civilize the Indians,
their only neighbors, and to teach them the rudiments of
agriculture so that they might have some means of sub-
sistence more certain than the precarious pursuit of wild

game. They hoped also that as the Indians became farmers they would become attached to the plat of land that furnished them a living, and would gradually give up the roving life that kept them always unsettled and savage.

When Astoria was founded ten potatoes were planted. These produced one hundred ninety. The next year the crop increased to five bushels, which yielded, in 1813, fifty bushels. When Mr. Parker came out in 1835 he brought with him a quart of seed wheat. This was even more prolific, for eleven years after its first planting it yielded a crop of twenty thousand thirty bushels. This was the striking result of the lessons in agriculture given by Whitman and Spalding. Under their earnest direction the object-lesson struck deep into the minds of the Indians, and many a brave of the Nez Perces and Cayuses looked with pride upon his little farm, though it must be admitted that he left most of the labor of cultivating it to his squaw. Yet with all of this it must be remembered that the Hudson's Bay people living at Vancouver had been successful to some extent in raising crops before the coming of the missionaries.

But the Hudson's Bay Company had aims exactly opposed to those of the missionaries. If the Indians turned to farming they would no longer bring rich store of furs to the Hudson's Bay posts to trade. Indeed, the American invasion threatened the very life of the Hudson's Bay business. Dr. Whitman, in his headstrong way, had insisted upon taking his wagon clear through to Oregon in spite of all the efforts of the Hudson's Bay factors to dissuade him. Never before had a wagon gone west of Fort Hall, and the authorities at that post pointed out that it would be sheer madness to make the attempt. But Whitman not only made the attempt, he succeeded. Where his wagon had gone others could follow, and the vision of long trains of American wagons loaded with American settlers became a nightmare to the Hudson's Bay officials. They caused magazine articles to be circulated in the East warning the people that Oregon was a desolate region wholly unfit for farming, or, as Daniel Webster put it, "Fit only for the prairie-dog and the Indian." The mongrel followers of the Hudson's Bay trade worked upon the suspicious minds of the Indians, arousing them to hostility against the whites, and even prejudicing them against the missionaries. At the same time the great monopoly was striving hard to get English settlers to come, preferring, if the country must be settled, to see it done by the British. But while we recognize all this, we must also remember that Dr. John McLoughlin, head of the Hudson's Bay interests in all that country, was a just and generous man, who never took a petty advantage and never failed to relieve the necessities of the American emigrants who came to his post at Fort Vancouver, destitute and miserable after their weary journey. He was kind to the Lees, to Parker, to Whitman himself. Indeed, but for him and his able lieutenants the massacre that put an end to Whitman's life and work would have been much more dreadful than it was.

While the few who had gone to Oregon were laboring manfully to prepare for future immigration, men of foresight in the East were working just as earnestly to arouse the public to a sense of the value of that great region. Most notable of these was Hall J. Kelley, a Boston schoolmaster, who had become inflamed with the idea that Americans would commit a crime against posterity should they let this region pass into British hands. With voice and pen he poured forth argument and appeal that his countrymen awake before the English had wrested Oregon from America forever. He even made a voyage to the far western coast, and came back more enthusiastic than ever. You see it was wholly a question as to who should first settle the country. In 1818 our government and England had made a treaty, agreeing that the citizens of both countries should be free to settle there. This was renewed in 1828. The Webster-Ashburton Treaty of 1842 settled the northeastern boundary between Maine and Canada, but left the question of the northwestern boundary still unsettled. Indeed, Webster

THE OREGON TRAIL (Western Wyoming)
Made deeper and wider by Whitman's Caravan.

seems not to have been awake to the importance of this great region, and there is every reason to fear that he would have accepted the Columbia River as a boundary had it not been for the opposition of able men like Senators Benton and Linn of Missouri, and the clamor of Hall J. Kelley and his followers.

The settlers in Oregon gradually became dissatisfied with the conditions surrounding them, and felt that our government, absorbed in other perplexing affairs of State, had forgotten the people who had gone west to help develop the territory along the western coast. More settlers were needed. The little farms were so far apart that the missionaries became lonesome and keenly felt their isolation and experienced a desire for additional people of their kind. The settlers were disappointed that no agreement had been arrived at by our country and England relative to the boundary lines. The only laws that governed the people were those put in force by the Hudson's Bay Company. Our laws did not extend out into that territory, and the Americans objected to the indifference of our government. Several attempts had been made to establish some sort of a government of their own but without success until 1843, when a provisional "compact" government was agreed to until such time as the United States should have authority over them. To this "compact," not unlike the "Mayflower Compact," we must accord the honor of being the first American government on the Pacific.

The missions started by Dr. Whitman and his associates on the upper Columbia had not met with the success that they deserved, and his Board of Missions did not feel that the results of the work being done in the territory warranted the continuance of the mission. Whitman was anxious to appear in person before this Board, and to urge it not to abandon the work. His wish to see the Board, and his desire to help the Oregon people determined the missionary to go east at once. Snow was in the mountains and it would be a trip full of danger, but he felt that he must start immediately and do what was in his power for the cause he represented and the settlers in the Oregon country.

The Indians had by this time become very hostile along the Oregon Trail. They were up in arms because they thought that the palefaces were coming in too great numbers and would gradually push the red man out of his hunting-grounds. As in the days of Lewis and Clark and the trappers, the Blackfeet and Sioux were the most dreaded. After thinking it over, Whitman determined to go by a southern route and thus avoid them.

When Whitman started from Walla Walla on the third of October, 1842, he took with him on his journey Amos Lovejoy, a recent immigrant to Oregon. These two men made the long and dangerous trip on horseback, in the middle of winter, through a country entirely new to them and beset with Indians. The first part of the journey was made over the Oregon Trail to Fort Hall; from there they pushed south, crossed the Colorado Mountains, forded the Green River, breaking their way through ice, crossed the mountain passes over ten thousand feet high where snow lay piled so deep that they had to break a way for the horses, camped out night after night in imminent danger of freezing, eating what little they could find, but always forging ahead until they reached Taos, the home of Kit Carson. At Bent's Fort Lovejoy had to abandon the trip, and from this point Whitman made the rest of the journey to the Missouri without guide or companion. By the third of March, 1843, he reached Washington, after having experienced every hardship imaginable through five months of continuous riding.

A strange and interesting appearance this missionary must have made upon the streets of Washington with his long hair, tanned skin, fur hat, buckskin trousers, moccasins, and leather coat, a typical backwoodsman. In his interview with the Secretary of War he learned that a treaty had just been signed by Daniel Webster and Lord Ashburton, envoy from England, but that the Oregon question was not mentioned, and that the status of that territory was still an unsettled matter between the United States and Great Britain.

Most of the information, which the statesmen at Washing-

THE GREEN RIVER PALISADES (Wyoming). A SCENE ON THE OVERLAND TRAIL

ton had received about that country of Oregon was from British sources, and naturally was not of a nature to inspire the authorities at Washington with a great desire to possess the territory. It was left to Whitman to present the facts as they appeared to him, setting forth the advantages of possessing the wonderful piece of agricultural home-making country on the Pacific coast. He advised the building of a series of forts along the Oregon Trail as a protection from the Indians, and farming stations to furnish supplies to the emigrants. He advised, in addition, that every encouragement be given to those seeking homes in the West.

From Washington Whitman went to Boston to endeavor to convince his Board of the necessity of continuing the missions in the Oregon country. When a complete explanation was given of the condition of the missions as they existed in Oregon, the authorities finally decided to continue the missions in that part of the country.

By June 1, 1843, Whitman was at Independence, ready to start once more for the West. At this point he identified himself with a colony, headed by Captain Peter H. Burnett, who afterwards became governor of California, bound for Oregon. This, the largest party that had gone west thus far, was made up of the best type of American citizens,—men, women, and children. They had two hundred wagons and over one thousand head of cattle. They went determined to stay, and took with them farm implements, seed grain, and many cherished pieces of furniture. The grain and tools they retained to the end; but little by little the furniture was burned or thrown down by the road as it became clear that they must lighten the load if they hoped to cover the many miles of rugged road that lay between them and Oregon. Whitman was ever present to hearten them in their despondency, to cheer them on to greater efforts, to remind them of the magnitude of their destiny. He it was that

persuaded them to cling to their wagons in spite of all difficulties, and that showed them how to take the wagons through
all the way to Oregon. Now, the Oregon Trail was really a
road. Never after this migration could it be claimed that
a wagon road to Oregon was impracticable. In the years
immediately succeeding still greater companies, all with
wagons and cattle, went over the road until the old trail
became worn so deep and wide that the Indians looking at
it would cover their mouths with their hands in awe, calling
it "The Great Medicine Road of the Whites," and declaring
that the rest of the country must be left empty, so many
people had gone to Oregon.

The party reached Oregon in October, 1843, and at once
commenced to build their houses. In the spring they plowed
their ground and planted their seed. The soil was rich, the
sun was bright, and plenty of rain came to the valley of the
Willamette, so that from the first it became a land of plenty.

Like many another brave soul that has labored in a great
cause, Dr. Whitman did not live to know of the final triumph.
The Indians killed him and his wife in November, 1847, and,
though the treaty settling the boundary between Canada
and the Oregon country at the 49th parallel had been signed
the preceding year, so slow was the transit of news in those
days that the glad tidings had not yet penetrated to the
mountain mission where the Whitmans worked so earnestly
to better the condition of those who were even then plotting
their destruction. Mrs. Whitman taught the school, in
which she had at one time five hundred Indian boys and
men, nursed the sick, and instructed the mothers and daugh-
ters in the rudiments of domestic economy. The doctor
labored to keep alive the interest in agriculture, preached
the lessons of Christianity, and healed the sick. But, alas!
he could not heal all the sick. Measles came, introduced
possibly by the family of some immigrant, and, although the

good doctor succeeded in curing the whites he could not cure
the Indians. There were two reasons for this: first, this disease
was much more fatal with Indians than with whites; then
again, the natives refused to carry out the instructions of the
doctor, for when the disease was at its height the fever-
parched Indians plunged
into the river to cool off.
There could be but one
result—death. But the
red men could make no
allowances. To their sus-
picious minds, inflamed
by renegade characters
among the half-breeds,
Dr. Whitman was in
league with the whites to
kill them and take their
lands. Further, the
Great Spirit was angry
because they had taken
another religion than
that of their fathers. So
they killed Whitman and
his wife and twelve im-
migrants in a massacre

PETER SKEENE OGDEN

that lasted for eight days. The Spaldings at their mission
of Lapwai were warned in time to escape to the shelter
of the nearest Hudson's Bay post. Peter Skeene Ogden,
the noted Hudson's Bay partisan after whom Ogden's
Hole in Utah was named, was sent to quiet the Indians and
to rescue the forty white people who had been captured. It
has been said that Ogden was the only person who could have
accomplished such a daring deed. This, for a time, was the
end of the Protestant missions in the eastern part of the

Oregon country.   Eventually Whitman College arose from the ashes of the Whitman mission.

The English had not succeeded in civilizing and settling Oregon.   America succeeded where England failed, because she brought to that country the family, the home, while the English came for adventure and for gain with traps, snares, and guns.   Seed wheat, corn, potatoes, and fruit trees proved the better materials for colonization.   The trapper was satisfied with newspapers coming by the dog-train route, six months old, but the settler had his printing-press with him.   One brought into that country all he had, the other took out of it all he could get.   The American's wagons were loaded when they entered Oregon, the English left that country each year with boats packed to their full capacity; one was continuously bringing in, the other constantly taking out; small wonder that one succeeded where the other failed.

4. FATHER DE SMET

When the chiefs of the Nez Perce and the Flathead tribes visited Captain Clark in the autumn of 1832, they really asked for Catholic missionaries, desiring the priest of "the black robe" to come to their country.   The reason for this was that some of the Indians of the Pacific waterways had been instructed in the Catholic religion by a few Christian Iroquois from Canada, who were in the service of the fur traders.

It was to these Indians that Father Jean Pierre De Smet was sent to carry the "White Man's Book."   De Smet was a Belgian by birth, but had come to the United States when yet a boy.   He was at his mission at Council Bluffs, working among the Potawatomi Indians, when the delegation of Flathead Indians passed on their way to St. Louis.   This visit inspired De Smet with such fervor that he asked for permission to go to the Rocky Mountains and investigate the condition of the Indians, with the possibility of establishing a mission among them.

In 1840 the good Father left Westport with a party of American Fur Company men. These men were on their way to the mountains where the Flatheads made their home. In this party were some thirty trappers, and an Indian named Ignace, who was to act as guide for De Smet to the home of the tribe. Pierre, another Indian, had gone ahead several months before to tell his people that the "Black Robe" would be at Green River in the spring.

De Smet's Travels
Courtesy of Lathrop C. Harper
FATHER DE SMET, TAKEN IN HIS YOUTH

This caravan also went over the Oregon Trail to the Green River rendezvous. Just before its arrival De Smet was met by ten of the most trusted warriors, who had been selected by the chief of the Flatheads. For a few days De Smet spent his time among the trappers and traders, and on his first Sunday in this region, the fifth of July, 1840, the Father celebrated Mass before the Indians, white men, traders, trappers, and hunters, a mixed, but a most attentive, congregation.

The altar for the celebration was placed on an elevation, and decorated with boughs from the cottonwood trees and fresh wild flowers of the plains. This sacred spot was known after this event as "la prairie de la Messe," the prairie of the Mass, so named by the Canadians of the camp. The devoted missionary first spoke in English, then French, and then, through an interpreter, to the Flatheads and Snake Indians, and the Canadians sang their hymns in French and Latin, the Indians joining in with their native tongues.

From Green River De Smet went to Pierre's Hole, where he found sixteen hundred Indians who had come through to meet him, some of them having journeyed eight hundred miles. Among these natives were Flatheads, Nez Perces, Pend d'Oreilles, and Kalispels. De Smet's entrance into the rendezvous was a triumphal march and his reception was royal. The chief, Tiolizhilzay (Big Face), ran to meet the missionary, followed by the men, women, and children, all eager to shake hands with the "Black Robe." When he met De Smet he exclaimed: "Black Robe, my heart was very glad when I learned who you were. Never has my lodge seen a greater day. As soon as I received the news of your coming, I had my big kettle filled to give you a feast in the midst of my warriors. Be welcome. I have had my best three dogs killed in your honor; they were very fat." The missionary testifies that the flesh of the wild dog was very delicate and extremely good, resembling the meat of a young pig.

The Father immediately began his work with the members of the tribes who assembled that evening in numbers amounting to two thousand. De Smet wept with joy that first night at the progress he had made among the natives, and for the opportunity that had been given to him to bring the tidings of salvation to the natives of the mountains and plains.

After spending two months with the Flatheads, going with them across the Divide, camping with them at the Three Forks, where thirty-four years before Lewis and Clark had camped, De Smet left his neophytes, and, accompanied by a select band of Indians to escort him past the hostile Blackfeet, pushed toward the northeast. From Fort Union De Smet departed for St. Louis, where he arrived December 31st, having made the entire journey in nine months.

Early the next spring, 1841, Father De Smet with two other priests and three laymen again returned to the moun-

Northern Pacific Railway
INTERIOR OF SAINT MARY'S MISSION. USED IN THE WORK OF FATHERS DE SMET AND RAVALLI. (Stevensville, Montana)

tains, where at South Pass the little band was met by ten lodges of the faithful Flatheads. After doing some missionary work in the southwestern part of Wyoming, De Smet and his men went to Fort Hall, where the British factor greeted the travelers with abounding hospitality. From here they were escorted by the Indians to what afterwards became Fort Owen in Montana, where they founded the mission of St. Mary's. From St. Mary's De Smet went on a long journey to Fort Vancouver, where Dr. McLoughlin met him with a most cordial greeting. It is interesting to note that while in this valley he also called on Dr. Whitman, presenting him with a Bible which is now preserved by the Oregon Historical Society.

Upon De Smet's return to St. Mary's he became so impressed with the prospect "for a harvest of souls" that he determined to go to Europe and obtain financial aid and additional assistants for his work. Returning by the way of the Yellowstone and the Missouri, he arrived again in St. Louis the last of October, 1842.

During Father De Smet's visit to Europe he engaged a number of Sisters of Notre Dame to come to America and help him build a convent and school in the Willamette Valley. With him also were one Belgian and three Italian priests. By the way of Cape Horn they all journeyed to the northwest coast, landing there in July, 1844.

After establishing the nuns in a convent, De Smet went over the mountains to his beloved Flatheads, where he performed prodigious labors. In addition to this tribe he was also determined to Christianize the dreaded Blackfeet, who were constantly making war upon his peaceful Indians. In this work with these Indians De Smet did not experience much success until the year 1846, when, on his way to St. Louis to obtain permission and aid to enlarge his missions, he visited the Blackfeet, spending three weeks with them. These fighting Blackfeet rather questioned the invasion of their territory by this strange, fearless white man with cloth of black and cross of gold. But his peaceful face and gentle manners were very reassuring. De Smet tells how the greatest chief hesitated about receiving him, when finally he extended his hand and invited the missionary to sit down on a strong and beautiful buffalo skin. Thinking that the pipe of peace was to be smoked from this robe, the Father took a seat in the center of it, when to his sudden surprise twelve of the chiefs took hold of the robe, and with De Smet in the middle carried robe and contents to a place some distance away where a successful council was held. This visit resulted in establishing peace between the Flathead chiefs who were with De Smet and the warring Blackfeet.

On his way down the Missouri De Smet met Brigham Young and the Mormons on their way to the West and the

De Smet's Travels.   Courtesy of Lathrop C. Harper

THE BLACKFEET SIOUX WELCOMING FATHER DE SMET

land of promise.   Of these people De Smet said: "They will one day probably form a prominent part of the history of the Far West." The missionary was a great prophet as well as a great priest.

This strenuous minister was not permitted to return and work among his "dear Flatheads," as his church had other work for him to do in St. Louis.   However, he occasionally made a visit for supervision of the work he had commenced, and which he had left other "Black Robes" to complete.

So great was De Smet's influence with these natives that

our government called him three times to help in pacifying them and assist in important negotiations. When the Indians with whom De Smet had worked grew ugly and out of humor, they would be immediately restored to order when a "Black Robe" appeared in their midst.

Northern Pacific Railway
OLD CHURCH BUILDING AT ST. IGNATIUS, INDIAN MISSION,
MONTANA, ON FLATHEAD RESERVATION

One story is told of Father De Smet which will illustrate his fearlessness, and, as a result of this, the faith which the natives had in the God-fearing missionary. The Crow Indians at first received De Smet with the awe and veneration that other tribes had given him; but after a time they became accustomed to seeing the black robe and the large gold cross, that he always wore on his breast, and the time came when they were skeptical of the powers which some Indians believed had been given him by the "Great Spirit." Finally, to test his spiritual power, one of the chiefs said that if De Smet would go and put his hand on the head of an

old wild buffalo bull that was grazing on the plains the
tribe would believe that he was a representative of this
"Great Spirit"; if he should be killed by the beast they would
know he was an imposter. This was not during the day of
miracles, yet De Smet knew he must make an attempt to

SAINT PETER'S MISSION, AN OLD INDIAN MISSION NEAR GREAT
FALLS, MONTANA

justify his calling. The grazing animal did not notice De
Smet's approach until the two were only a few yards apart,
when the big creature raised his head and looked at the black
gown and the glittering gold cross, but he did not move.
Going nearer and nearer slowly and quietly, De Smet finally
placed his hand on the buffalo's head. When the priest
walked back to the tribe he was received with additional
awe and reverence as one coming from the "Great Spirit."
This deed was heralded not only throughout the Crow tribe,
but to all of the Northwest, where De Smet's power was
believed to be God-given.

As late as 1868 Father De Smet visited the mountain region, going to Cheyenne, Wyoming. While he was there he told the people who conversed with him of the wonderful amount of gold that was to be found in the Rockies, and of the great future there was for the people who were to live in that region. All who knew Father De Smet spoke of him in the highest terms, regretting that he gave up the missionary work in which he had done much to bring civilization to the Northwest.

5. THE MORMONS IN UTAH

The other religious movements toward the West were made in the hope and belief that the Indians might be Christianized; the religious organization known as The Church of Jesus Christ of Latter-Day Saints came to the great West in order to establish a new home for the followers of its faith. The desire for freedom in thought and freedom in worship made this entire sect abandon their homes east of the Mississippi, and move out into the valley of the great inland sea.

The Mormons, as the Latter-Day Saints are called, were organized in the state of New York in 1830 by Joseph Smith, thence they moved to Ohio, and soon to Missouri. From this state they were driven to Illinois, and upon the recurrence of mob violence here, in which their prophet Joseph Smith was killed, they decided to cast their lot in the wilderness, away from conflicting authority, where they might establish a community of their own.

In 1847 the great exodus to the West was made under the guidance of Brigham Young, the president of the Mormon church and successor of Joseph Smith. The first stage of the journey was made in 1846 through Iowa. General Stephen W. Kearny, then commandant at Fort Leavenworth, was eager to play a worthy part in the Mexican War and was straining every nerve to recruit his force to the necessary strength. Five hundred of the Mormons joined his standard. His force thus weakened, Brigham Young did not attempt to cross the vast, Indian-swept plains that fall, but camped on the banks of the Missouri until spring.

LION HOUSE AND BEEHIVE HOUSE, RESIDENCES OF BRIGHAM
YOUNG. SALT LAKE CITY
The high wall is now replaced by an iron fence.

These Mormons who gave their services to the cause of the war became known as the Mormon Battalion, and went first as far south as Santa Fé. Here a large number of the men became too ill to render service in the army and were sent to Pueblo for rest and treatment; from here they ultimately made their way to the Great Salt Lake Basin. The rest of the battalion went to San Diego under the command of Brigadier-General Cooke, and some of these men also returned to Salt Lake by the way of the California trail, while a number of them stayed in California and one of their number is said to have been the first man to discover gold there.

When the warm days in April came, in 1847, the emigrant band that had camped on the banks of the Missouri started on the journey for their new home in the Rocky Mountains,— the exact location of which no one knew.

Brigham Young had read very carefully the reports that

IN ECHO CAÑON, UTAH

Fremont had made about his explorations in the West; he had talked with Father De Smet about agricultural lands toward the Pacific; and had interviewed every possible person who had been in the wilderness. That there was a home for him and his people somewhere toward the setting sun no one of the band doubted.

The Mormon trail was from Council Bluffs to the Platte, along the north side of the Platte to Fort Laramie, thence to Ft. Bridger over the Oregon Trail, and southwest to Great Salt Lake. As other Mormons were to follow this first band, many devices were adopted to guide succeeding parties and to give them news of the pioneers. Letters were stuck in the skulls of buffalo found on the prairies, and for guideposts they painted on the space between the horns of the skulls the date of their arrival at that spot.

The Oregon Trail was now growing wider, for the Mormons averaged only thirteen miles a day in order to give their cattle plenty of time to graze on each side of the beaten trail. Broad and deep was this highway, and broader and deeper did it become, until it became a wide belt of furrows. Rain, snow, wind, and time have not been able unto this day to obliterate the tracks of the Indian, trapper, trader, explorer, missionary, settler, soldier, freighter, stage-driver, and express-rider. Just before the Mormons reached Salt Lake Valley, Bridger met them and told Brigham Young all he knew about that region. Bridger laughed at the idea that any crops could grow in that desert and swore that he would give one hundred dollars for the first ear of corn that was raised there. The quiet answer was: "Wait a little and we will show you." And the world has been shown what a resolute, determined, and able-bodied set of men, led by an able leader, may accomplish.

On July 24, 1847, the little caravan trailed down into the "land of promise." From the sagebrush desert found there has evolved a land that from the tops of the hills looks like a checkerboard, with every square a beautiful farm or orchard. The desert, by the means of irrigating ditches, was turned into one grand garden where many blades of grass grow where none grew before.

After the first detachment of Mormons reached the valley others came, and then others year after year, until in all parts of Utah were rich communities. Brigham Young did not permit all of his followers to huddle in the valley of the Great Salt Lake, but sent small communities here and there

SALT LAKE CITY AND THE WASATCH MOUNTAINS

to make settlements in the wilderness. So we find Logan in the rich Cache Valley where the old trappers used to hold their rendezvous, and Ogden in Ogden's Hole, named by the trappers after old Peter Ogden, a trapper for the Hudson's Bay Company. In the valley of Utah Lake, to the south, is Provo, perpetuating the name of Etienne Provost, the partisan of Ashley, the fur man.

MONUMENT TO BRIGHAM YOUNG AND THE PIONEERS, SALT LAKE CITY

To this day the 24th of July is celebrated throughout Utah with speeches, music, picnics, races, and processions; for it was on this day in 1847 that Brigham Young and his band of one hundred forty-three men, women, and children passed through Emigration Cañon and went down into the great valley. And on this July day each year, Pioneer Day, the people of Utah rig up queer old wagons with tattered, weather-beaten covers, drawn by a cow and a horse, escorted by roughly dressed men, and march in procession with old women and young children, dogs, cows, and mules, a ragged lot, in commemoration of the great march of their early pioneers.

REFERENCES

Lummis. The Spanish Pioneers.

Royce. History of California.

Dellenbaugh. Breaking the Wilderness.

Thwaites. Early Western Travels, Wyeth, Townsend and De Smet Letters.

Dye. McLoughlin and Old Oregon.

Irving. Captain Bonneville.

Eells. Marcus Whitman.

Dye. McDonald of Oregon.

Barrow. History of Oregon.

Schafer. History of the Pacific Northwest.

Lyman. The Columbia River.

Bancroft. History of Utah.

Inman. The Great Salt Lake Trail.

Stansbury. Explorations.

Coutant. History of Wyoming.

Whitney. History of Utah.

Meany. History of the State of Washington.

Paxson. The Last American Frontier.

CHAPTER V

FREMONT'S EXPLORATIONS

1. THE WIND RIVER MOUNTAIN EXPLORATION, 1842-43

John Charles Fremont made five journeys of western exploration; three of these were under the direction and pay of the United States government; two of them were private ventures; all of them were made between the years of 1842 and 1854.

It might be profitable to go back a few years and learn for what reasons Fremont made these explorations, and why he was selected to make them.

There lived in St. Louis at this time Thomas H. Benton, United States Senator for thirty years, having been the first senator to be sent from the then new state of Missouri, in 1821. Most naturally the development of the new West became of vital importance to Benton, and it was to him that the authorities at Washington turned for the most accurate information of the wilderness. Captain William Clark was connected by marriage with the family of Benton, and a natural intimacy existed between these two men,—explorer and statesman.

Captain Clark had continued in his office of superintendent of all of the Western Indians ever since his return from the West in 1806. His influence was necessarily very power-

ful with the Indians as well as with the authorities at
Washington, particularly the Indian Bureau and the
Department of the Interior.   As a matter of fact, Clark

JOHN CHARLES FREMONT

made most of the
treaties with the
Indians during his
term of office.

The constant com-
panionship with this
explorer inspired
Senator Benton with
the idea of learning
accurately through
government explora-
tions the geography
and topography of
this vast region, that
treaties might be
made, railroad proj-
ects acted upon, and
settlement directed
more wisely than
could be the case
with the hearsay
information then
available. Other
people were taken
into this family council.   The fur traders came, smarting
from the arrogance of the Hudson's Bay Company.   Chief
among these fur men were the Chouteaus, an old French
family who had carried on their fur trade for sixty years.
Then there were the picturesque Mexican merchants; the
military men from the frontier, fresh from their skirmishes
with the Indians; the "Black Robes," who added their bits

of information of the mountains and the prairies; the French voyageurs; and lastly, the wealthy merchants from Spain, France, and America who were interested in their trade that stretched across Mexico to the "Sea of Cortez," as the Gulf of California was then called. All of these men brought their different information and expressed their views as to the best course to pursue to better conditions in the great West. Benton became absorbed in the idea of obtaining information from an official source that might be published by the government and distributed among the people. Believing in this method of spreading knowledge about the new West, Benton carried his plans to Washington, where he had very great influence.

Young Fremont had been in the service of the government for some time as a member of the Topographical Corps, which had to make surveys, plans and estimates for proposed routes for canals and roads to be used for commercial and military purposes. In 1840 he came to St. Louis from one of his northwest geological surveys, and while there met Senator Benton. He also met Jessie Benton, daughter of the Senator, who became Mrs. Fremont in 1841 while Fremont was a lieutenant in the United States army. In the following May this army officer, under the directions and instructions of our government, left St. Louis to explore the country lying between the Missouri and the Rocky Mountains, going by way of the Kansas and Platte rivers.

By steamboat Fremont went from St. Louis as far as the mouth of the Kansas, where his final preparations were made at the trading-post of Chouteau about ten miles up the river from "Kansas Landing." In the caravan were twenty-one enlisted men, a topographical engineer, Charles Preuss, a hunter, Lucien Maxwell, and a guide, Kit Carson. In addition to this escort there were Henry Benton, a young man of nineteen years, son of the Senator, and Brant, a boy twelve years old, who was sent on the expedition "for the development of mind and body which such an expedition would give."

Following on, and parallel to, the Santa Fé Trail for over one day, the expedition traveled to the northwest until the Platte was reached, where a majority of the men went directly to Fort Laramie by the way of the north branch, while Fremont went up the South Fork, entered Wyoming about thirty miles southeast of Cheyenne, and then pushed north until he came to the fort where the other division had pitched its tents. These men had overtaken Jim Bridger, who caused them much alarm by telling them that the Sioux, who were on the warpath, had sworn to make war upon every living thing that might be found west of Red Buttes, a point on the proposed path of the explorers. The authorities at the fort also warned them of the dangerous mood of the Indians, and advised the explorer to wait until the warring spirit of the hostile tribes had been subdued. Even the chiefs, Otter Hat, Breaker of Arrows, Black Night, and Bull's Trail, all insisted that there was serious danger ahead for the explorers.

Nevertheless, Fremont determined to push forward and meet the enemies if necessary. He left at the fort part of the baggage, field-notes, and records of observations, as well as young Benton and Brant, as it was thought that these boys were too young to encounter the expected dangers. Chronometers, thermometers, transits, and barometers were easily transported to South Pass, where more Indians were encountered, who urged the explorer to return to the fort for protection.

After reaching the rift in the mountains, Fremont's chief purpose was to climb the highest peak that shone out in its white, glittering, silvery splendor to the northwest. As the party approached the mountains food became scarce, the

daily supply consisting of buffalo meat fried in tallow.
For the last climb a few pounds of coffee and a small quantity
of macaroni were carefully saved.

Fifteen sore-footed mules and fourteen level-headed men
made up the mountain party which traveled for two days
toward the coveted peak. Here Fremont, with sextant,
spyglass, compass, barometer, and two men, made the final

FREMONT ADDRESSING THE INDIANS AT FORT LARAMIE

climb on through deep defiles in the mountains, past steep,
rocky and slippery places, where each side was a perpendicular wall of granite three thousand feet above their heads.
Finally they picketed the mules, determined to go the rest
of the way on foot. The ascent was necessarily slow, as they
now were so high that it was difficult to breathe, the violent
exercise affecting the action of their hearts.

Fremont now exchanged his heavy moccasins for ones with
very thin soles, so that toes as well as fingers might assist in
scaling the almost perpendicular rocks. Often both hands
and feet had to be put into crevices to get over the rocks,
until with a final spring Fremont was on the top where there
was a sloping rock about three feet wide.

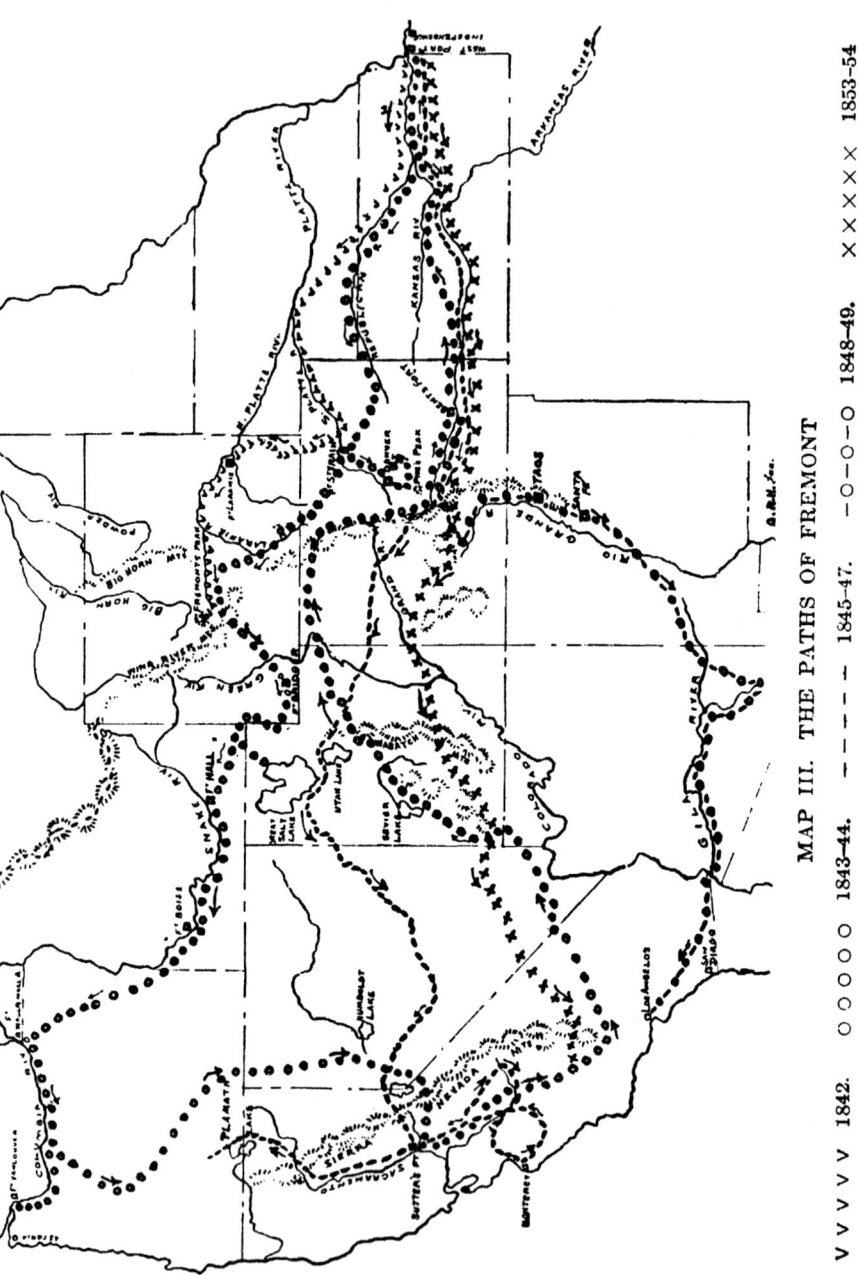

MAP III. THE PATHS OF FREMONT

VVVVV 1842. OOOOO 1843-44. - - - - - 1845-47. -O-O-O 1848-49. XXXXX 1853-54

After Fremont let himself down from the rock, each of the men in turn climbed to the top on the "unstable and precarious slab." The barometer, as it was placed in the snow at the summit, registered 13,570 feet higher than the waters of the ocean. Putting a ramrod in a crevice, the party then unfurled the Stars and Stripes. For a few moments the men sat in the awful silence and terrible solitude. The only sign of life they had seen that day had been a small sparrowlike bird, but "while sitting on the rock a solitary bee came winging his flight from the eastern valley and lit on the knee of one of the men." The bee was captured and put between the leaves of a notebook with flowers that had been collected during the day. From this high point these indomitable explorers saw hundreds of small lakes, the source of the Green, which emptied into the Gulf of California, after changing its name to the Colorado; on another side was the Wind River valley, in which were the headwaters of the Yellowstone, a branch of the Missouri; to the north were the Tetons with their white caps, furnishing water to both the Missouri and the Columbia; and away to the southeast were the peaks whose snows supplied the water for the Platte.

The descent did not take long. The mules were easily found and the camp was soon reached, where Kit Carson, who had been sent back with the extra men and mules not needed in the last climb to the mountain top, had everything comfortable for them.

The peak discovered bears the name of the man who first climbed it, and is situated in a county of that name in the center of the State of Wyoming.

Fremont's return journey was down the cañons, where the river had worked its way through the mountains not far from Independence Rock. On this rock Fremont engraved with a hard piece of granite a "symbol of Christian faith."

On this "rock of the Far West" he traced many other names of those who had gone over the trail before him. When Fremont is called "The Pathfinder" the historians make a mistake, for in truth he only found the path that had been made by others, Stuart, Ashley, Bonneville, Wyeth, Bridger, Carson, Lee, Whitman, and hundreds of pioneer path-breakers. Fremont, "The Mapmaker," would be a much more accurate title, and at the same time do more justice to the real trail-makers.

After joining the rest of the party, the explorers returned to St. Louis by way of the Platte, arriving at their destination about the middle of October. The next day Fremont started for Washington, D. C., where he spent the winter preparing a report of the expedition. Preuss made the maps, representing each day's journey, and our government finally issued the whole report. In order to further assist the travelers there was indicated on the maps where camps could be made and where grass and water might be found.

Congress had many thousands of these reports printed and distributed. The greatest interest was aroused not by this report alone but by the subsequent reports by Fremont as well, all of which were used as handbooks and guides by western travelers. All who had their eyes turned westward — the Oregon enthusiasts, and the thousands they were inciting to move, the Mormons smarting under intolerable conditions — all read these reports. In the mind of the general public Fremont was the one man that knew the West. It is no wonder that they named him "The Pathfinder."

2. THE GREAT SALT LAKE, THE COLUMBIA AND CALIFORNIA EXPEDITION, 1843–44.

Fremont's second expedition was organized by the government in order to examine the large territory south of the Columbia River lying between the Rockies and the Pacific. The agitation over the Oregon country was the direct reason for the second venture. Our government needed a road to the mouth of the Columbia River.

Fremont met his men at the little town of Kansas, now Kansas City, and formed the caravan, on May 29, 1843, to go up the valley of the Kansas River to the headwaters of the Arkansas. To map out a new road to Oregon and California in a milder climate than the northern trails encountered was the first object of the expedition.

Knowing that he was to be among Indians who were noted for treachery as well as bravery, Fremont applied to Colonel Kearny for a brass twelve-pound howitzer, which was furnished from the arsenal at St. Louis. The party consisted of Creoles and Canadians, French and Americans, making in all thirty-nine men.

Thomas Fitzpatrick, called "Broken Hand," one of his hands having been shattered by a gunshot, was selected as the guide. Preuss, the topographer, was also a member of the expedition, as was Basil Lajeunesse, Fremont's favorite on this journey.

When the party reached Pueblo, Kit Carson, who happened to be in that part of the country, was persuaded to join the expedition. From St. Vrain's Fort on the South Platte Fremont pushed westward up the Cache la Poudre, in Colorado, to the Big Laramie in Wyoming, by the way of the Medicine Bow Range, on to the North Platte, and northwest to the Sweetwater and to South Pass. This route from Fort St. Vrain (Colorado) into the Laramie plains

KIT CARSON

was practically a new one. Others had been over it, but the Indians were hostile along its line and it was not used. It was over this path that Jacques La Ramie was traveling when he met his death. This trail as surveyed by Fremont was a practical one, over which in a few years hundreds of emigrants made their way to the West. Later on a stage line was established along this route.

From South Pass the expedition went by the way of Green

River to the Bear, and down that stream to the Great Salt Lake, arriving there in September, 1843. The day after Fremont arrived at the lake he took his rubber boat, not unlike the one in which he went down the cañon of the "Upper Great Platte," and with Carson, Preuss, and the Frenchmen set out to explore the lake. This trip Fremont said was "the first ever attempted by white men on the lake." This, of course, was an error, as it is known that James Bridger and other trappers had been on the lake in their bull-skin boats many years before.

VIEW OF GREAT SALT LAKE FROM HAT ISLAND

Fremont and his men rowed in their leaking air-blown boat to an island, where they finally landed with much difficulty. This, now called Fremont Island, they named Disappointment Island, because they found no game or grass there. The boat was so unsafe that they attempted no further exploration of the lake.

After the expedition left Great Salt Lake it went directly to Fort Hall, and the reunited party then traveled north and west until it reached The Dalles. Thus was completed the purpose of Fremont's second expedition, the uniting of his survey with that of Captain Wilkes, who was then surveying on the Columbia. The connecting of the two surveys made a continuous exploration from the Missouri to the Pacific.

From The Dalles Fremont took a boat trip to Vancouver, and was most courteously and hospitably received by Dr. McLoughlin, who always made a "forcible and delightful impression on a traveler from the long wilderness."

After Fremont's return to The Dalles, he started out to explore the Sierras and the western part of the Great Basin. No one of the expedition was familiar with the route, the

A VIEW ON HAT ISLAND (VISITED BY FREMONT) IN GREAT SALT LAKE

general supposition being that there was a great river, the Buenaventura, that flowed westward from the region of the lake through the Sierra Nevadas and into the Pacific. This supposition was a mistake, and Bonneville's expeditions had shown its fallacy, but Fremont seems not to have been familiar with Bonneville's work.

The expedition traveled southward until it came to Tlamath (Klamath) Lake, from which three streams departed, one to the western ocean, one to the Columbia, and one south to California. After crossing the eastern slope of the Sierra Nevadas the party continued southward.

About this time Indians of a new tribe made their appearance, and Fremont attempted to obtain from them a guide

to pilot the expedition over the mountains. For the services
of the guide he offered many presents of bright-colored cloth
and showy trinkets. The Indians conferred with each other,
but pointed to the snow on the mountains and "drew their
hands across their necks, and raised them above their heads
to show the depth, signifying that it was impossible to get
through." However, the Indians
directed the expedition to go farther
south, where a pass in the mountains would be found, after passing
through which white men were to
be found at the end of two days'
travel. For this journey to the
south a guide was finally furnished,
after many gifts were presented to
him.

The guide was a young Indian
dressed in most scanty clothing.
He suffered intensely, for by the
time the pass was reached it was
snowing hard, and the weather had

PELICANS ON HAT ISLAND,
GREAT SALT LAKE

turned extremely cold. In order that the guide might not
desert at this important time, Fremont had him march
between two guards each armed with a rifle, for the poor
fellow showed signs of panic as the snow came down on his
naked skin. After a while this Indian was allowed to return
to his people, and another was selected who was at once
properly clothed in leggins, mackinaws, and a large blanket
in addition to the bright red and blue cloth which had been
presented to him for his services and which he wore as a
decoration of honor. Now the dreaded time had come
when the supplies of food had entirely given out and the
camp dog had to supply the soup-pot; but it made a
"strengthening meal."

The snow was heavy and deep, and a road had to be broken through it over the Sierra. In order to do this sledges and snowshoes had to be made for the scouting party, which marched in single file to the top of the mountains. From here the men saw to the west, just outlined in the distance, a large valley without snow, and beyond this a low range of mountains which Carson recognized at once with delight as the mountains of the Pacific coast,—"It is fifteen years since I saw it, but I am just as sure as if I had seen it yesterday."

Between these mountains and the summit on which they stood was the Sacramento Valley. Back again the scouting party went to the camp, which was twenty miles distant, in order to bring the animals and the baggage over the mountains. This was a very difficult task, as the animals would break through the snow, and the sun shining on the snow made the men nearly blind. After pulling and tugging and cutting down trees to put in the path for the animals to walk on, the summit was finally reached February 20, 1844, this time with all of the baggage and the animals. The scientific instruments showed that the exploration party was now one thousand miles from The Dalles, two thousand feet higher than South Pass.

On the day after Fremont had crossed the summit he observed a line of water to the west directing its course to a larger body of water. This told the explorers that they now were looking at the Sacramento Valley and the Bay of California.

With Carson as a guide, Fremont pushed ahead to find Sutter's Fort,[1] as provisions of all kinds had given out and the party was in a most deplorable condition. Mules and horses had to be killed for food, roots and wild onions, and even the leather of their saddles were eagerly chewed by

[1] See Chapter VI.

the starving men. So extreme was the exposure and anxiety that two of the men became insane. Preuss lost the trail and had nothing with him with which to dig the wild onion except a pocket-knife; but he found some ants and some frogs, which kept him alive until he met some Indians who gave him plenty of roasted acorns and mussels, and directed him to the path of his comrades.

When the explorers reached Fort Sutter they were met by the genial captain, who gave them a frank and cordial reception. In the valley of the Sacramento at the junction of the American and Sacramento rivers the party camped. The trip across the mountains had been especially hard on the animals. When Fremont started to cross the summit he had sixty-seven mules and horses; when he reached the valley there were but thirty-three of them, and these were in such forlorn condition that they had to be led.

The explorers stayed in the valley only two weeks, during which time all preparations were made for the home-going trip. They collected horses, mules, and cattle; Sutter's mill went night and day to grind them flour, and toward the last · of March the expedition started on its eastern trip, having left six of its company in California.

The journey home was first toward the south, then around the southern extremity of the Sierra, and northward on the Old Spanish Trail to Utah Lake. Thus the explorers completed a circuit of three thousand five hundred miles between September, 1843, and May of the next year. From Utah Lake, Fremont went east, explored the Colorado River, the headwaters of the Arkansas, North, Middle, and South Parks (Colorado), and went to Bent's Fort on the Santa Fé Trail, where the party was saluted with a display of the American flag and the firing of guns. At this fort Carson and three of the men ended their connection with the explorers, since all danger was over and there was no further need of a guide. After arriving at St. Louis the expedition disbanded and the members scattered. Fremont went to Washington, where he wrote up the report of this second expedition. This was printed with the first report, and ten thousand extra copies were made for distribution throughout the country.

For services rendered on this second expedition Fremont was appointed by President Tyler captain by brevet "for gallant and highly meritorious services in two expeditions."

Fremont was enthusiastic over California, its climate, its vegetation, and its unusual commercial position. After his return he had two earnest wishes: one, to see California settled with Americans, and the other, to make that beautiful country his home.

3. THE MEXICAN WAR, 1845-47

The third government expedition under the direction of Fremont reached Bent's Fort by the way of the Arkansas on August 2, 1845. Here Fremont again engaged Carson and Fitzpatrick to act as guides, and with sixty other experienced men started for the Pacific. The object of this exploration was to follow up the Arkansas to its source in the Rocky Mountains, to complete the survey of the Great Salt Lake, and to extend this survey southwest to the Sierra Nevada and Coast Range, and to determine the best route by which to reach the Pacific.

With no difficulty the party reached the southern end of the Great Salt Lake, where the men surveyed and explored for two weeks. The desert west of the lake was an unknown district to all; but Carson and two of the men traveled for sixty miles to the west, sending up signals of smoke to guide the rest. At last they crossed the Sierras and came to Sutter's Fort.

For many years trappers had brought to the States glowing reports of the beautiful and rich California, which at that time belonged to Mexico. Indeed, so many Americans were settling in that country and Texas, bringing with them their ideas of independence, that Mexico passed an act in 1833 forbidding foreign colonization in her border provinces.

But the law scarcely checked the flow of American colonists to these outlying territories, particularly California, which had always proved attractive to every one that knew it. In 1846 there were in California about four hundred Americans out of a population of nine thousand. Most of these Americans were at Monterey, the great center of trade, and in the Sacramento Valley, where the settlers had their ranches in the neighborhood of Sutter's Fort.

When the Mexican authorities in California heard of Fremont's entrance into the Sacramento Valley they became alarmed and questioned the right of the invasion. Sutter explained to the authorities that the expedition had only come to make some surveys in order to ascertain the best route from the United States to the Pacific Ocean; that the trip was made in the interest of science and commerce; and that the men composing the party were citizens, and not soldiers.

Finally, after Fremont had been to Monterey to explain in person to General Castro, the governor, why he was in California, and to obtain his permission to explore the country, the explorers made a permanent camp in the valley of San José. This incensed the Mexicans, and General Castro revoked the permission to explore, ordering Fremont to withdraw from California, and stating that Americans were not allowed to settle in that province under the law of 1833.

Fremont and his men then went north into the Oregon country. On his way he had serious conflicts with the

Indians of the Klamath tribe, who killed Basil Lajeunesse and three other men. Fremont had scarcely arrived in the Oregon country when he received a message that caused him to hasten back to California. War had been declared between Mexico and the United States.

When Fremont again reached California, May, 1846, he found the Americans much alarmed over the reports that Castro was organizing an army to drive them out of California. The American ranch people under command of Fremont seized the town of Sonoma and hauled down the Mexican flag on which was a representation of a grizzly bear. This was the commencement of what was known as the "Bear Flag Insurrection." Following this came an armed conflict between the Californians and the United States. Fremont, acting as an officer under the command of Commodore Robert Stockton, took a prominent part in the uprising when Monterey, Los Angeles, San Diego, and many other cities surrendered to the American forces.

It now became necessary to apprise our government of the condition of affairs in California. On September 5, 1846, Carson started on the long trip to Washington to carry despatches from Fremont. Carson thought that he would be able by fast and continuous traveling to make the journey in two months. Going by the way of the Gila Trail, by October 6th he was east of the Rockies, not many miles from his home at Taos. Here he was met by General Kearny, who was on the march to occupy California. Carson told Kearny that he was too late, that California had been conquered, but the General ordered the guide to go back with him, sending the despatches to Washington by another man. The Gila Trail, being the most practical road for horses, was again put into use. By December they reached California, where they found that the Mexicans had not been conquered by any means, and where Kearny himself came near being conquered in a severe combat in which one half of his men were killed and the rest cooped up on a barren hill. Relief must be had at once to save Kearny from, surrender, and Kit Carson, accompanied only by Lieutenant Beale of the army, after a journey of extreme peril and hardship, found Stockton at San Diego in time to save brave Kearny and his men.

Carson joined Fremont at Los Angeles, and stood by him all through the unfortunate dispute that disgraced our conquest of California. Stockton and Kearny each claimed chief command there, and Fremont sided with Stockton of the navy rather than with Kearny of the army. For this he was court-martialed and left the service in disgrace. Thus ended in gloom and misfortune this third expedition that began so gloriously.

4. THE FIRST PRIVATE VENTURE, 1848-49

The fourth expedition of Fremont was a private venture, undertaken in 1848 by him and Senator Benton, with the intent to show the world that Fremont was still a great leader who deserved better treatment than his country had given him. Fremont's purpose was to demonstrate the shortest and best route to California. Final arrangements were made from Fort Bent, when the expedition started up the Arkansas, November 25th, with every hope of being able to cross the San Juan Mountains, a branch of the Rocky Mountain system, without great difficulty or hardship. It was Fremont's chief misfortune that Carson was not with him, and that he had to trust to Bill Williams to lead him through the labyrinth of cañons west of the Arkansas. Williams was not equal to the task. After leaving Wet Mountain Valley and forging through Robideaux's Pass they became hopelessly lost in the mountains

MONUMENT TO KIT CARSON IN SANTA FÉ.
NEW MEXICO

of Colorado. Winter came unusually early, storms of unwonted severity obliterated the trails and filled the cañons with snow; food supplies became exhausted; men and animals died; and had it not been for Fremont's pluck and endurance none would have returned to civilization. The starving condition of his men determined Fremont to attempt to find Taos down the Rio Grande, where food and horses might be obtained. Following the stream, he finally found Carson at his home in Taos, and the two headed a relief party to rescue what was left of the expedition. The suffering they had endured would be difficult of description. With the ground covered with heavy snow, no grass or shrubs to be obtained, the mules resorted to eating the blankets that were put on them at night, as well as the blankets of the men. The men were reduced to the utmost extreme — that of cannibalism. Of the hundred mules and horses every one died, and eleven of the thirty-two men left their bones in the Colorado mountains. Evidently Fremont was not one of Fortune's favorites. But, as unconquerable as ever, he took the expedition through from Taos to California by the way of Santa Fé, Albuquerque, and the Gila Trail.

5. THE LAST EXPEDITION, 1853-54

Fremont was one of the men who will not admit failure. Like Frederick the Great, he was a small man with a great spirit, and like Frederick, too, he fought the harder when Fate was most unkind. At his own expense he fitted out another expedition in 1853, determined to accomplish what he had so utterly failed to do in 1848. He took but twenty men, half of them Delaware Indians. The starting-point of this, as of the previous expeditions, was Independence. Final arrangements were made at Fort Bent, and the real experience commenced in Colorado where the Rocky Mountains were crossed at Cochetopa Pass, not far from the scene of the terrible suffering of the previous expedition. Winter was fast approaching, and much precious time was lost by the party in searching for passes while floundering through deep snow that had by this time commenced to fill up ravines and obliterate any trace of a possible path. On this journey the prostrations were numerous and the suffering beyond description, though the Sierras, not the Rockies, proved to be the real land of trouble. The horses had to be killed for food, and when things came to the worst Fremont called his men together and made them promise that no matter how great their suffering they would not resort to cannibalism as the men of the fourth expedition had done. Fremont manfully said to his starving men: "If we are to die, let us die together like men." Then it was that the white men, Indians, and Mexicans, in the darkness of the night, amidst the deep snow on the mountain top, with a zero wind chilling the very marrow of their bones, entered into this solemn compact which was faithfully kept. Yet, so great was their suffering that the men were forced to eat cactus, the leather of their saddles and even the hide and burned bones of the horses whose flesh had long since been consumed. Thus they lived for fifty days, tramping through the snow with Fremont leading the way and breaking the path.

While on the Green River, before crossing the Wasatch Mountains, an alarm of "Indians" was given, and at the same time sixty mounted Utahs, all with rifles or bows and arrows, came bearing down on the explorers' camp, threatening immediate extermination. Fremont with his usual composure gave a Colt's Navy six-shooter to one of his men and told him to shoot at a small piece of paper that had been torn from his record book and fastened to a tree. The instructions were to fire at intervals of from ten to fifteen seconds to call the attention of the Indians to the fact that it was not necessary for white men to reload their arms. When the first shot was fired, and the paper squarely hit, it made no further impression on the natives than for them to point to their rifles, as much as to say, "Yes, we can do that." But when the second shot went off without a change in the position of the arm they were much startled; when the third shot came they were not only startled but curious and confused. The revolver was then handed to one of the chiefs who fired and hit the paper; the fifth and sixth shots were made by two other Indians. By this time the natives were thoroughly frightened into acknowledging that they were at the mercy of the white man who could shoot his gun without loading it, and they calmly submitted to the requests of the explorer and his men.

Finally, after the Green River was passed and the guide had gone astray, Fremont took the course that had been described to him by the mountain men and found passes in the mountains all of the way to California. This route lay along a line between the 38th and 39th parallels, running through Colorado, Utah, and Nevada.

"Something of the practical value of these explorations may be inferred from the fact that the great railroads connecting the West and the East lie in a large measure through the country explored by Fremont, and sometimes in the very lines he followed."

REFERENCES

Prince.   Historical Sketches of New Mexico.
Semple.   American History and its Geographical Conditions.
Bancroft.   History of California, Arizona and New Mexico.
Bruce.   The Romance of American Expansion.
Inman.   The Santa Fé Trail.
Inman.   The Great Salt Lake Trail.
Grinnell.   Trails of the Pathfinders.
Fremont.   Memoirs of My Life.
Coutant.   History of Wyoming.
Hough.   The Way to the West.
Dellenbaugh.   Breaking the Wilderness.
Century Magazine, Vol. XLI.
Royce.   History of California.

CHAPTER VI

THE GOLD DISCOVERIES

1. CALIFORNIA

In closing an earlier chapter we said that the fur trade was not only the most romantic but also the most important factor in the early history of the Great West. It is equally true that the gold discoveries hold first place in its later history, both for romance and for significance. Gold was first discovered in California at Sutter's Fort, the post so often mentioned in the course of this narrative. The fort was built near the junction of the Sacramento and American rivers, by Captain John A. Sutter, a Swiss gentleman who came to Oregon with the early pioneers from Missouri, and soon moved to California, where both he and his fort became famous. Emigrants, as early as 1846, began to enter the Sacramento Valley, always going to Sutter's Fort for supplies and horses. Realizing that much grain would be needed for flour, Captain Sutter instructed his carpenter, James W. Marshall, to erect a sawmill to make lumber for a flour-mill. In order to turn the wheel of this mill, a dam was constructed, the water from which ran in a channel, or race. In this race in 1848 Marshall first found little specks of bright-colored gravel, some of them as large as a grain of wheat. Being skeptical of their worth, Marshall sought the assistance of Sutter and an old encyclopedia. Every test that was made confirmed

156

the belief that gold had been found in paying quantities. Much precaution was taken to keep the matter a secret, at least until the mill was built, for fear that the men would abandon their work in order to hunt for the precious metal. But such a secret could not long be kept. The men all deserted, and an examination of the ravines and creeks showed that gold was everywhere. The people living outside of the valley were skeptical of the richness of the find at first, but the continued reports brought people into the valley from San Francisco, San José, Monterey, and down as far south as San Diego. Soldiers and sailors deserted, men left their farms, towns were depopulated. The fever for gold soon became epidemic. "The whole country from San Francisco to Los Angeles, and from seashore to the base of the Sierra Nevada resounded to the sordid cry gold! gold! gold!" The report of the gold to be found in the streams was carried to the States by the Mormons who had helped build the sawmill, and who afterwards went to Salt Lake. These were some of the men who served in the Mormon Battalion and had gone to California by the way of the Gila Trail.

The outgoing ships from San Francisco had taken the news to other parts of the world. As a result an invasion of goldseekers came from the Hawaiian Islands, from Oregon, from Mexico, and from far-away China. It is well to remember that the first gold mined in California, and indeed in all the great gold camps of the West, was free gold, lying loose in the sand of the streams, the gravel of old creek beds, and the gullies of hillsides. The particles varied in size from the tiniest dots, called "scales," to nuggets so valuable as to stagger belief. Some of these nuggets of pure gold found were worth $3,000. At $18 an ounce this represents almost twelve pounds of pure gold in a single lump. These finds were rare, however, and most miners had to content themselves with taking $20 to $100 a day out of the sand, in the form of

countless particles, called gold dust or, in the short verna-
cular of the day, "dust." They needed none of the expensive machinery used to-day, when most of our gold is mined
in the form of ore that is embedded in rocks that must be
crushed in a great mill before the mercury and cyanide used
to separate the gold from the rock can be brought into play.
In those days a man who had a good claim could make big
wages with no other tools than a pickaxe, shovel, and tin pan.
Those who operated on a large scale had no more expensive
apparatus than a series of wooden troughs, called sluice-
boxes. The gold-bearing gravel would be thrown into the
upper one of these, and a stream of water would be rushed
over it. The gravel, being comparatively light, would be
carried away and deposited as "tailings" at the end of the
lowest sluice-box, while the gold, being heavy, would sink
and be held fast by the mercury which lined the bottom of
the box, and which attracts and holds gold much as a magnet
attracts steel. To be sure, such a crude affair could not
save all the gold. Many of the old tailing dumps are still
so rich in it that men are making comfortable wages in
working them over. This washing out of free gold is called
placer mining, and the ease with which it could be carried
on explains why all sorts of men stampeded to the placer
fields of the West.

The amount of the metal found by each man of course
varied very much. Some panned out $1,000 a day, and
occasionally the amount would reach $5,000 for a single day's
labor. It is estimated that during the year 1848 gold to
the amount of $5,000,000 was taken out of the streams of
Sacramento Valley.

We are told that at no time since the discovery of the New
World by Columbus, when gold and silver went in such
abundance to Europe, had there been such a wide-spread interest in the finding of gold. Every ship that could be put

into use took the gold-seekers to California. "From Maine to Texas the noise of preparation for travel was heard in every town. The name of California was in every mouth; it was the current theme for conversation, song, and sermon. Every scrap of information concerning the country was eagerly devoured."

There were three routes that were taken by the people bound for the new gold fields: First, the sea voyage around South America; second, the sea and land journey by way of the Isthmus of Panama; and third, overland, in wagons, on horseback or on foot. By the overland ways there were two chief routes that lay along the trails: First, on the Santa Fé Trail to the city of that name, and thence over the old Spanish road or over the Gila Trail; second, the northern route, over the old Oregon, Great Salt Lake, and California trails. This northern route was the one most used by the gold-seekers, who were, as we know, following in the paths of the Indians, trappers, and explorers.

Again, the starting-place for the new El Dorado was from Independence and St. Joseph, although some of the parties left the Missouri at Council Bluffs. Vehicles of all descriptions now traveled on the trails, for the path had become pretty well worn and easy of travel. There was the prairie schooner with its white canvas tent-like cover, drawn by oxen, and the two-wheeled cart with the old family horse. "Ho, for the diggings!" was not an unusual sign to be seen painted on the canvas of the wagons.

In the month of April, 1849, twenty thousand people left the Missouri River for the gold fields, and "by the summer there was a continuous caravan from Independence to Fort Laramie."[1] One traveler over the trail said that he counted four hundred fifty-nine wagons within the space of nine miles. Stansbury says: "The road was literally strewn with arti-

[1] Bancroft.

cles that had been thrown away. Bar-iron and steel, large blacksmith anvils, bellows, crowbars, drills, augers, gold washers, chisels, axes, lead, trunks, spades, plows, grindstones, baking ovens, cooking stoves without number, kegs, barrels, harness, clothing, bacon, and beans were found along the road in pretty much the order enumerated.''

While some families were trying to get rid of their possessions others clung to theirs until the end of the journey. An interesting description is given of a Dutchman who drove six yoke of oxen with a heavy wagon loaded to the top with household furniture of every description. Behind him came the wife, driving a covered wagon in which were numerous children. On the back of the wagon was fastened a large chicken-coop filled with hens, close after these came the milch cows, followed by the old, gentle, worn-out family horse on the back of which sat a browned girl, while in the extreme rear trotted the growing colt.

The journey from Fort Laramie was not a hard one as far as Fort Hall, or Salt Lake. It was when the desert was reached west of the lake that real danger began. In fact, many of the weary people did not attempt to go farther than Salt Lake and rested with the Mormons until spring.

The scarcity of water in the desert caused the most suffering. After traveling along the banks of Humboldt River for many miles the traveler found that the water disappeared in a hole or "sink," and he was left unprepared for the barren tract of land that lay beyond to the southwest.

It is estimated that during the first year of the "forty-niners" ninety thousand people went to California. We must not suppose that these all were poor emigrants seeking a home, for every class of people, in all callings and professions, was fully represented. This gold fever was, also, not limited by any means to the inhabitants of the United States, for all races, colors, and conditions of men flocked to this new country, and, as if by night, the valley along the Sierra grew into a city of tents.

Prices for everything rose to unheard of heights. No one thought of using change to an amount less than fifty cents. A dollar was the price of a newspaper, and even at that price it was eagerly purchased even though a month old.

The gold produced in 1848 was equal to $5,000,000, while in 1849 it jumped to $23,000,000, and the fame of California spread far and wide, attracting the gambler, the border-ruffian, and the criminal, who all made conditions dangerous and often unbearable. The "vigilance committees" of the miners took matters into their own hands, and the offenders were punished under the terms of lynch law.

The lust for gold created new centers of population. Towns of tents grew into cities of substantial houses. San Francisco, with a population of a few hundred before the discovery of gold, became a large commercial center with fifty thousand inhabitants in 1860. Stockton and Sacramento developed rapidly, becoming interior supply stations, where multitudes of people flocked to purchase necessities. Salt Lake City benefited by the exodus from the States to the gold field, for not only did the emigrants purchase from the Mormons grain and needed vegetables, but thousands decided to go no farther, and to make their home in the fertile valley of these successful farmers. By 1850 over 10,000 people had settled in the region of Great Salt Lake.

People often ask, "What did Marshall and Sutter gain from the gold strike in the Sacramento Valley?" Nothing. Both of these men expected to make large fortunes from the sawmill. But, inside of a few months, all of the large and desirable trees had been cut down by the miners, and the wheels of the mill no longer turned. Sutter had large tracts of land in the valley, but the squatters and lawyers managed to take them all from him.

Both of these men for a time received a pension, but Marshall "at the age of seventy-three died alone in a solitary cabin. He was buried at Coloma in sight of the place where he discovered the gold. His figure, in colossal bronze, stands over his grave."

2. NEVADA

It has wisely been said that "the history of the Comstock lode is to a great extent the history of Nevada." The Mormons had established trading-posts along the trail of the California miners, particularly in Carson Valley, where paying dirt had attracted many prospectors. In this valley Carson City was founded in 1858, but the real excitement came the following year with the discovery of silver near Gold Hill, situated but a few miles east of Lake Tahoe. This was the famous Comstock lode, which not only attracted the world by its marvelous wealth but made Virginia, Carson, and Gold Hill cities. No other mining excitement in this country ever equaled the wild and widespread mania for gold and silver in Nevada. It must be remembered that Nevada was at that time the western part of Utah, inhabited by the native Digger Indians and a few Mormons.

The same class of people that rushed to California poured into this new mining camp. Within five years something like $100,000,000 worth of gold and silver were taken from the sides of the mountains, and of this amount only about one-third was gold. Nevada, when she was admitted as a state into the Union, was known as the "Silver State," on account of the preponderance of the native silver found in her hillsides. Since 1861 this Comstock lode has yielded $350,000,000 of bullion. Of this 40% was gold and 60% silver.

Marvelous veins of gold and silver were discovered in many other places in this new ore field; the richness of the finds made the world gold-mad, and the tide of fortune-seekers crowded the trails. In Mark Twain's "Roughing It" you will find a graphic picture of these wild days. He was there during the most exciting time, living in Virginia City and reporting on the "Territorial Enterprise."

If the precious metals had not been discovered in the mountains of the Pacific coast, the West would still be to-day in an undeveloped and unsettled condition. A large part of this floating population stayed in the West to "grow up with the country," and made their homes amidst the mountains and valleys and in this manner completed another chapter in western development. Nevada has experienced in recent years another exciting period. The mining camps of Tonapah and Goldfield have become prosperous towns with an annual output of gold and silver amounting to many millions of dollars. At Ely, copper is produced on a large scale. Modern methods of mining and the erection of smelters in the mining districts have made many of the old abandoned mines very valuable.

3. COLORADO

When Pike was captive in Mexico, Pursley showed the explorer a shot-pouch of nuggets which he had found on the head of the Platte River. It is a strange coincidence that after half a century gold should be found in fabulous quantities in the home of Pursley's nuggets, and that Pike's Peak should be the center of the early treasure-sought district.

Many of the early trappers carried nuggets in their shot-pouches which they asserted they had found in the mountains, but furs paid better and no particular attention was paid to the gold until California sent the reports of her rich fields to the people of the East. It is true also that the Indians knew of the gold-bearing mountains, as did the missionaries who were bringing the gospel to the natives. Father De Smet knew of the valuable gold deposits in the Rocky Mountains, but dreaded to have the facts known for fear that miners might occupy the country and exterminate his beloved Indians. It has been said that the early missionaries of California knew of the rich metal in the Sacramento Valley,

DENVER IN 1865

but they also feared the invasion of the white men before their good work could be completed.

In 1858 a party of Cherokee Indians, who had been in California looking for gold and lands on which to make their homes, discovered that gold existed in the sands of Cherry Creek, Colorado, and other streams of that region. After returning to their home in the southwest, they organized a mining party and returned to the Rocky Mountains to explore and dig for gold. From this beginning numerous parties were formed that year to work in the mountains of Colorado. It was during this year that Colorado Springs and Colorado City were laid out. Then, within five miles of the present site of Denver, a city of twenty cabins was started, which was called "Montana." In September, St. Charles was built at the mouth of Cherry Creek, and a month later a little settlement on the west bank of the creek was called "Auraria." The three towns ultimately united and were known as Denver.

In the summer of 1859 there were as many as 150,000 gold-seekers within the present boundaries of Colorado, but one-third of these soon turned back toward the States, completely disgusted, swearing that the reports of riches in the Pike's Peak region were all lies. Their trail, too, was strewn with household goods of every description, with many a broken wagon and worn-out horse, and too often with the bones of the adventurers. But even as they went homeward news came that their more persevering comrades had made tremendous strikes, and away they went for the mountains again pell-mell,—so hard it is to resist the lure of gold.

From now on mining camps sprang up in every direction, and thousands of emigrants came to the new gold fields. People fairly pushed each other in order to get to the mountains first, wild to make a big strike. One day there would be a rumor of a discovery, and the people would swarm to that locality, "alighting like locusts upon a field which could not furnish ground for one in a thousand of those who came. Finding themselves too late, they swarmed again at some other spot, which they abandoned in a similar manner."

Cities and towns grew up in a day, and flourishing little settlements dotted the gulches and the ravines. These mining camps developed into the towns of Golden, Central City, Golden Gate, Black Hawk, and Georgetown.

A KIT CARSON AND EARLY PIONEER MONUMENT ERECTED IN
DENVER, 1911

4. MONTANA

When Lewis and Clark passed over the land now within the boundaries of the state of Montana they were utterly unconscious of the hidden fields of gold that were often under their feet. Had the truth been known, the men of the expedition never would have reached the coast, for the party would

Northern Pacific Railway
PIONEER GULCH. ONE OF THE FIRST PLACER MINING SPOTS
IN MONTANA

have disbanded at once and gone to digging. Had the trappers and fur men discovered the gold, how different would have been the early development of that territory! Without doubt, it would have come into its own long before California or Oregon territory.

For years, centuries it may be, these regions were the haunts of the Indians. Then again for a long time this district was the trapping-ground of the fur men, and was un-

Northern Pacific Railway
AN OLD-TIME ARRASTRE USED IN GRINDING GOLD AND SILVER ORE

From Northern Pacific Railway
VIRGINIA CITY, MONTANA, IN THE SIXTIES

molested by the western tramp of the home-seeker, or overland traveler. But the time came when the gulches and
mountains could no longer conceal their secret, and gold-mad
people claimed the territory.

It was in 1862, fifty-seven years after Lewis and Clark had
passed through the country, that John White and William
Eads found "pay dirt" on Grasshopper Creek, and the town
of Bannock sprang up there with mining-camp rapidity and
for a short time was the capital of Montana.

In 1863 rich gold deposits were found in Alder Gulch,
where Virginia City, first called Varina, sprang into existence,
reaching a population of over four thousand inside of a year.
Virginia City was the seat of government until 1875, when it
was removed to Helena. The tide of immigration during
these years divided into three streams: one flowing to southern Idaho and Oregon, the second to California, and the
third to Montana. Helena, in 1864, became a rival of

Northern Pacific Railway
THE FIRST BOARD HOUSE ERECTED IN HELENA, MONTANA

Virginia City, owing to the rich discoveries made by John Cowan in Last Chance Gulch. During the first season over $16,000,000 of gold was shipped out of that region. The fact that such fabulous sums of gold were being sent for transportation down the Missouri River attracted many desperate characters to the mining camps. It has been said that the population of the camp was divided into three classes: the disgusted Colorado miners, who were called the "Pike's Peakers," the emigrants from the East, and the disappointed miners from California. Among those who came there were many lawless desperadoes, fugitives, outlaws, and thieves who are always to be found in a frontier mining camp. This class of men kept the vigilance committees very busy. Many of these desperate men had a secret organization, by which they banded together and called themselves "road-agents." The chief object of this "agency" was to relieve the travelers of their gold dust or what valuables they might have. They were nothing more nor less than the old-fashioned highway robbers. They made life in the mountains terrible for a time, but finally their impudence became unbearable, the better citizens organized into vigilance committees to combat them, as had been done in California, and after a short period of desperate struggle the forces of law and order won.

5. IDAHO

The first discovery of gold in Idaho was made by E. D. Pierce in 1860 in the Clearwater country, and the mining camp of Pierce City was named after him. During the summer of this year rich deposits were found on Oro Fino Creek, where the tent city of Orofino was established. This placer mining region was above Lewiston, where the first real town was built, and which became Idaho's first capital. The real history of Idaho begins with the finding of gold in Boise

Basin in August, 1862. The Bannock Indians had known for many years of the yellow metal that lay hidden in the mountains. One day one of this tribe, watching Moses Splawn wash out these glittering particles of metal from the sand in the sluices, told the white man of a basin in the mountains where he had picked up "chunks" of this metal when he was a boy. Following the direction which the native gave of the country, Splawn finally found the gold, and here the famous Boise Basin Mining Camp suddenly grew into existence. By the next spring the usual stampede had commenced in earnest for the Idaho mines. Men from the California and Nevada camps tumbled over each other in their frenzied desire to arrive first; the agriculturists of Oregon and Washington added to the rush; and the gold-seekers from the East swelled the population.

The Indians became very hostile, and murdered many of the emigrants who were on their way to the camp. An appeal was made to the United States government, and a military fort was established on the Boise River near the present site of Boise about forty miles above old Fort Boise, the trading-post of the Hudson's Bay Fur Company.

During the year of 1863 thirty thousand emigrants came to southern Idaho, and with them rapid improvements and corresponding high prices. One merchant said: "I sold shovels at $12 apiece as fast as I could count them out." Another merchant brought a wagonload of cats and chickens to the "diggins." The chickens sold for $5 each and the cats sold for $10. Hay was sold as high as forty cents a pound, and was hard to obtain even at that rate. To buy a horse, one had to part with a knife, two blankets, a good shirt, a pair of leggins, a pocket mirror, and numerous small trinkets. If you will look carefully at your map of Idaho you will find Sinker Creek. The story is told that some men fishing on that stream and looking for something to serve as a weight for their line, fastened a gold nugget to the cord in place of the usual lead sinker. This simply illustrates the abundance of the yellow metal found during those exciting days.

6. THE FREIGHT, EXPRESS AND STAGE LINES, AND THE PONY EXPRESS

Freight. All these towns springing up in the mountains needed supplies. Their sole business was the production of gold. Everything they ate or wore or read, even the kerosene oil they used for making light and the tools they needed for digging gold or the scales they must have for weighing the dust had to be brought from the East. Hence sprang up a great freight business between the Missouri River and the mining camps in the mountains. Lummis in his "Pioneer Transportation in America" says that in 1860 five hundred freight wagons frequently passed Fort Kearney in a single day, that in one day 888 west-bound wagons, drawn by 10,650 oxen, mules, and horses, were counted between Fort Kearney and Julesburg, and that in 1865 six thousand wagons passed Fort Kearney in a period of six weeks. The single firm of Russell, Majors, and Waddell used at one time 6,250 freight wagons and 75,000 oxen. Probably there are not to-day so many oxen working in the United States as this one firm used in this western freight traffic half a century ago. It was a colossal business — this of supplying the necessities of the towns that had sprung up in the Far West, and it deepened and widened the Oregon Trail into a great highway, the like of which was never known before.

It took all summer for one of these freight trains to go to Bannock or Salt Lake and return. The danger of loss by fire, flood, or Indian attack was great. No wonder, then, that prices were high in those western towns, that a pound of flour sometimes cost a dollar and a gallon of kerosene a dollar and a half. No wonder that life was simple and the people learned to live without many of the things that had been deemed necessities "back home." To be sure, most of the freight that reached the mining camps of California, Nevada, and Idaho came from the west through the port of San Francisco, or up the Columbia to Umatilla, whence it was freighted by wagon to Pierce City, Orofino, Idaho City, and Boise; to be sure, also, very much freight went by boat up the Missouri to Fort Benton, and by a good road only 140 miles long to Helena, whence it was sent to Deer Lodge, Bannock, and Virginia City; still, all the towns in Colorado, Wyoming, and Utah depended upon the freight trains from Atchison, St. Joseph, and Kansas City to supply them. Even the goods that came by way of San Francisco or the Columbia had to be shipped around Cape Horn, a tedious and expensive process. In those days the Missouri River was a real highway and one of the chief commercial lines of the country. Its turbid waters witnessed an activity never seen there before or since. Almost every day boats left St. Louis or St. Joseph with passengers and freight bound for the Far West. Almost every day boats tied up at Fort Benton and discharged their cargoes of people and goods for distribution to the gold diggings. Fort Benton was a busy place then, and Helena, though great as a gold mining camp, was still greater as an emporium for the distribution of goods. It was this favoring geographic situation that enabled Helena to outstrip all the other towns of the territory, and to wrest the seat of government away from Virginia City.

The Stage Coach and Express. But the freight team was too slow for the transportation of passengers, express, and mail. Both East and West demanded more speedy service. To meet this demand the stage coach appeared. Some of these ancient vehicles may still be seen in western towns serenely decaying in the obscurity of back yards. They were jaunty enough in their day, with their cavernous bodies extended behind into a platform or "boot" for the reception of baggage, and were built high up in front to furnish a throne for the driver, who needed a high seat, not only that he might keep a better lookout for Indians and road-agents, but also that he might the better supervise the six horses bounding along under his skillful management. They were not uncomfortable, those old coaches, for the body swung on great leather straps, which softened the jolts and gave a gentle, swaying motion to the heavy contrivance. But of course even this comparative comfort became irksome after several weeks of constant travel, continuous through the hours of darkness as well as of daylight. It took two weeks to reach Helena from Atchison, and the fare was $150. It took three weeks to reach Sacramento. Do you begin to realize now what a boon railroads were to the West? It was a standing joke in California that the term of a member of Congress might expire before he ever got to Washington unless he had good luck.

With the help of the United States government a line of regular stages was established, and began to run in 1858. This was the southern route, better known as the "Butterfield Route," from the name of its founder. It ran from St. Louis to San Francisco by way of El Paso, Yuma, and Los Angeles. By going so far south it avoided the deep snows in the mountains of the north, but it increased the distance by seven hundred miles. This route was discontinued when the Civil War broke out, and the central route, known to the people of the West as Ben Holladay's Stage Line, was established in its stead. This line began to run in 1861, and held the field until superseded by the steam cars in 1869. It ran from St. Joseph, Missouri, the western terminus of the eastern railroads, to Sacramento, California, the eastern terminus of the Pacific coast railroads. This was a run of

1,900 miles and was made in eighteen days, if everything went well. We are already familiar with the names of the chief stations on this route, Fort Kearney, Fort Laramie, Fort Bridger, Salt Lake City, Carson City, and Placerville. You see it followed the Oregon Trail.

The government paid Holladay a million dollars a year for carrying mail. Add to this the passenger fares collected, and you will readily see that the gross receipts were a princely sum annually. But the famous Ben had to keep the line equipped with 100 expensive Concord coaches, 2,700 horses and mules, $55,000 worth of harness, and 250 men of more than ordinary skill and courage. Hay and oats cost him a million dollars a year, for he sometimes had to pay at the rate of $125 a ton for hay. At one time, in a spirit of western rivalry, he drove from Salt Lake to Atchison in eight days, but in doing it he ruined horses and equipment worth twenty thousand dollars. When, besides all this, we remember that the Indians stole his stock, killed his drivers, and burned his stations we can readily understand why Ben failed to get rich and why he was glad to sell out to the Wells, Fargo Company in 1866.

The rival stage line that aroused Ben to that frenzied race which cost him twenty thousand dollars was owned by Russell, Majors, and Waddell, the great freighting firm. They ran from Leavenworth, Kansas, to Salt Lake City by way of Denver and thence northwest over the trail mapped by Fremont in 1843. Over this route Majors drove from Salt Lake to Leavenworth, a distance of 1,200 miles, in ten days. A record which was a nightmare to Holladay till he beat it.

The Pony Express. Even the stages were not swift enough to suit those who were awaiting letters. To satisfy them the Pony Express was instituted to carry letters only. There is no more picturesque achievement of the plains than the operations of the Pony Express, carrying letters from Independ-

ence to Sacramento, a distance of 1,950 miles. "Never before
or since has mail been carried so fast, so far, and so long mere-
ly by horse power."[1]   The Pony Express was in operation
for about two years after April, 1860, during which time all
letters between the Missouri and the Pacific were carried in

SOUTH END OF GREAT SALT LAKE

small leather bags attached to the saddles of the daring riders.
This fast mail service went over the overland stage route
from the Missouri, across Kansas to the Platte, through the
Continental Divide by the way of South Pass, down Emigration Cañon to Salt Lake City; then, skirting the southern
shore of the lake, the road led across the desert to the Hum-

[1] Lummis.

boldt, along its waters into the desert at Carson Lake, into the Sierra Nevada Mountains, through a pass at the head of Carson River and by way of Placerville to Sacramento.

To keep the service in operation, five hundred enduring and fast ponies were in constant use, eighty expert riders and two hundred station men. In addition, one hundred and ninety stations, five to one hundred miles apart, were strung all along the line. Russell, Majors, and Waddell, running at this time a daily stage-coach line, undertook the management of this mail service, utilizing many of their stage stations for the express service. While the horses were picked with the greatest care, the selection of riders received even greater consideration. They must be wiry, cool, watchful, and quick. "It was no easy duty; horse and human flesh were strained to the limits of physical tension. Day or night, in sunshine or in storm, under the darkest skies, in the pale moonlight, and with only the stars at times to guide him, the brave rider must speed on. Rain, hail, snow, or sleet, there was no delay; his precious burden of letters demanded his best efforts under the stern necessities of the hazardous service; it brooked no detention; on he must ride. Sometimes his pathway led across level prairies, straight as the flight of an arrow. It was oftener a zigzag trail hugging the brink of awful precipices, and dark, narrow, cañons infested with watchful savages, eager for the scalp of the daring man who had the temerity to enter their mountain fastness." [1] When a rider arrived at one of the stations, he found his new mount saddled and bridled, and he was given just two minutes to change his horse and the mail, but he did it in less time, and was on, and off again almost before his foaming horse had come to a standstill. Two hundred fifty miles a day was the scheduled rate, and

[1] Inman. The Great Salt Lake Trail. By permission of Crane & Co., Topeka.

no surplus weight in rider or equipment was permitted, hence lithe young men were selected, and only a knife and a revolver allowed for self-defense. The mail-pouches were water-proof, sealed, and securely fastened to the front and back of the saddle. The letters were carefully wrapped in oilcloth and sealed. These pouches were locked at the starting-station and never opened until they reached the end of their route. The rate of postage was five dollars a half ounce, in the early months of the service, but this was later reduced to one dollar. All papers for the Pacific were printed on tissue paper, and went as letter postage in letter envelopes. Never but once on this long dangerous route was a mail-bag lost. One rider was scalped by the Indians, but the riderless pony came panting into the next station with the mail safely fast-ened to the empty saddle. The scheduled time for crossing the plains with this nimble California mustang service was ten days, but the last message of Buchanan went from St. Joseph to Sacramento in eight days and a few hours. The greatest achievement was accomplished when the news of Lincoln's inauguration was rushed over the two thousand miles in seven days and seventeen hours.

When William Cody, Buffalo Bill, was fourteen years of age he became a pony-express rider. Many were his hair-breadth escapes, and thrilling were his experiences as he rode back and forth over the trail. "While engaged in the ex-press service, his route lay between Red Buttes and the Three Crossings of the Sweetwater. It was a most dangerous, long, and lonely trail, with perilous crossings of swollen and turbu-lent streams. An average of fifteen miles an hour had to be made, including change of horses, detours for safety, and time for meals. Once, upon reaching Three Crossings, he found that the rider on the next division had been killed the night before, and he was called on to make the extra trip until another rider could be produced. This was a request compliance with which would involve the most taxing labors and an endurance few people are capable of; nevertheless young Cody was promptly on hand for the additional journey, and reached Rocky Ridge, the limit of the second route, on time. Then he rode back to Red Buttes without a rest. This round trip of three hundred twenty-one miles was made without a stop, except for meals and to change horses, and every station on the route was entered on time. This is one of the longest and best ridden pony-express journeys ever made, the entire distance being covered in twenty-one hours and thirty minutes."[1]

"Pony Bob" (Robert H. Haslam), who was on the first relay and in the service to its end, made a record ride of three hundred eight miles without leaving the saddle. The Indians had killed the man at the next station, and he passed not only the burning ruins of that station but two others before he found some one to take his place. For this terrible work these men received only $125 a month and board.

Mark Twain, who at one time staged along the Overland Trail, has very graphically described the pony-express rider, whom all in the stage had a great desire to see. "Presently the driver exclaims: 'Here he comes.' Every neck is stretched farther and every eye strained wider. Away across the endless dead level of the prairie a black speck appears against the sky, and it is plain that it moves. Well, I should think so! In a second or two it becomes a horse and a rider, rising and falling, rising and falling, sweeping toward us, nearer and nearer, growing more and more distinct, more and more sharply defined. Nearer and still nearer, and the flutter of the hoofs comes faintly to the ear. Another instant a whoop and a hurrah from our upper deck, a wave of the

[1] Majors. Seventy Years on the Frontier.

rider's hand but no reply, and the man and the horse burst past our excited faces and so winged away like a belated fragment of a storm." [1]

REFERENCES

Bancroft.　History of California.
Century Magazine, Vol. XLI.
Royce.　History of California.
Stansbury.　Report of Great Salt Lake.
McMurray.　Pioneers of the Rocky Mountains and the West.
Wheeler.　The Trail of Lewis and Clark.
Bancroft.　History of Nevada, Colorado and Montana.
Hailey.　History of Idaho.
Langford.　Vigilante Days and Ways.
Twain.　Roughing It.
Majors.　Seventy Years on the Frontier.
Inman.　The Great Salt Lake Trail,
Vischer.　The Pony Express.
Fairbanks.　The Western United States.
Cody.　Tales of the Plains.
Root and Connelley.　The Overland Stage to California.
Parrish.　The Great Plains.
Lummis.　Pioneer Transportation in America.
Whitney.　History of Utah.
Paxson.　The Last American Frontier.

[1] Roughing It.

CHAPTER VII

THE SOLDIER AND THE SETTLER

1. THE BOZEMAN TRAIL

The Soldier and the Settler came hand in hand to the West. The soldiers were not sent into the wilderness to tame the red man until the white man showed his intention to build his home in the land beyond the Missouri. The white man could not bring his family to the western plains and the Rocky Mountains, among the fierce Indians, until some protection was given by an armed force of men. The time for the actual settler had now arrived. The explorer, the fur hunter, the trader, the missionary, and the miner had each done his work in the direct preparation for the most important man of all, the actual settler.

Military occupation of the territory of the Indian and fur trapper now began. The government built Fort Logan on the Arkansas, near the site-of the historic Bent's Fort on the old Santa Fé Trail. The more celebrated Fort Union of the American Fur Company, at the mouth of the Yellowstone, was also obtained for a government post, as was the famous Fort Laramie on the Oregon Trail. The Santa Fé Trail was the military road from Fort Leavenworth to Fort Logan; the Oregon Trail, from Fort Leavenworth to Fort Laramie; the Bozeman Road led to the Indian country of northern Wyoming and Montana; and the South Platte Trail from Fort Bent, to Fort Laramie and then north by way of the Black Hills to Fort Pierre on the Missouri. These and other good fur trails became military roads without having to be made or altered at the expense of the government.

Then again the steamboats, introduced by the fur men on the Missouri, were ready to transport military troops and

Northern Pacific Railway
DIAMOND CITY AND CONFEDERATE GULCH IN THE SIXTIES
(Montana)

their stores, as well as the actual settler. Without these pioneer steps in the expansion of our western country the soldier would have been confronted with an impossible problem.

After gold was discovered in Montana there was need of a road from the Oregon Trail to that country for the convenience of the miners and the settlers who were then headed toward the land of precious metal. To assist in this migration, the Bozeman Road was mapped out. This route was not a main highway like some of the other trails, but a branch road, running from Red Buttes on the Platte to Three Forks on the Missouri, going by the way of Fort Laramie north through Wyoming and west through Montana through Bozeman Pass. The road was named for John M. Bozeman, a frontiersman who came to Montana in the early years of the sixties and first traced the trail, and who was killed by Indians in the Yellowstone Valley. The pretty town of Bozeman, Montana, Bozeman Creek, and Bozeman Pass, the one used by Clark on his homeward journey, also perpetuate his memory. Long before this road was constructed the Crow Indians very bitterly resented the invasion by white men of their country, in which were plenty of antelope, buffalo and deer, wild berries, and grass for their horses. All of the earlier travelers spoke in the most glowing terms of this beautiful land. It was not to be wondered at that the Indians hated to see any first step taken that would in the end deprive them of their ancestral hunting-ground. The Crow chief, Arapooish, told the government authorities that his tribe would not peacefully permit any invasion upon their land, and would war with the white man if he tried to inhabit their territory.

The Crows had already lost some of these lands to the Sioux tribe. For seventy years these Crow stoutly defended their lands from the Sioux, but finally by force of numbers the Sioux swept them out of all the choicer parts and sent them to skulk in the mountains or to seek refuge among their relatives, the Assiniboines of Canada. The Sioux, in turn, warned the whites that they would resist to the death any attempt to make a highway of these fine hunting-grounds, and it was found necessary to establish military posts to protect travel over the Bozeman Road. Fort Reno was established on Crazy Woman's Fork of Powder River, north of this, Fort Phil. Kearney was built on Piney

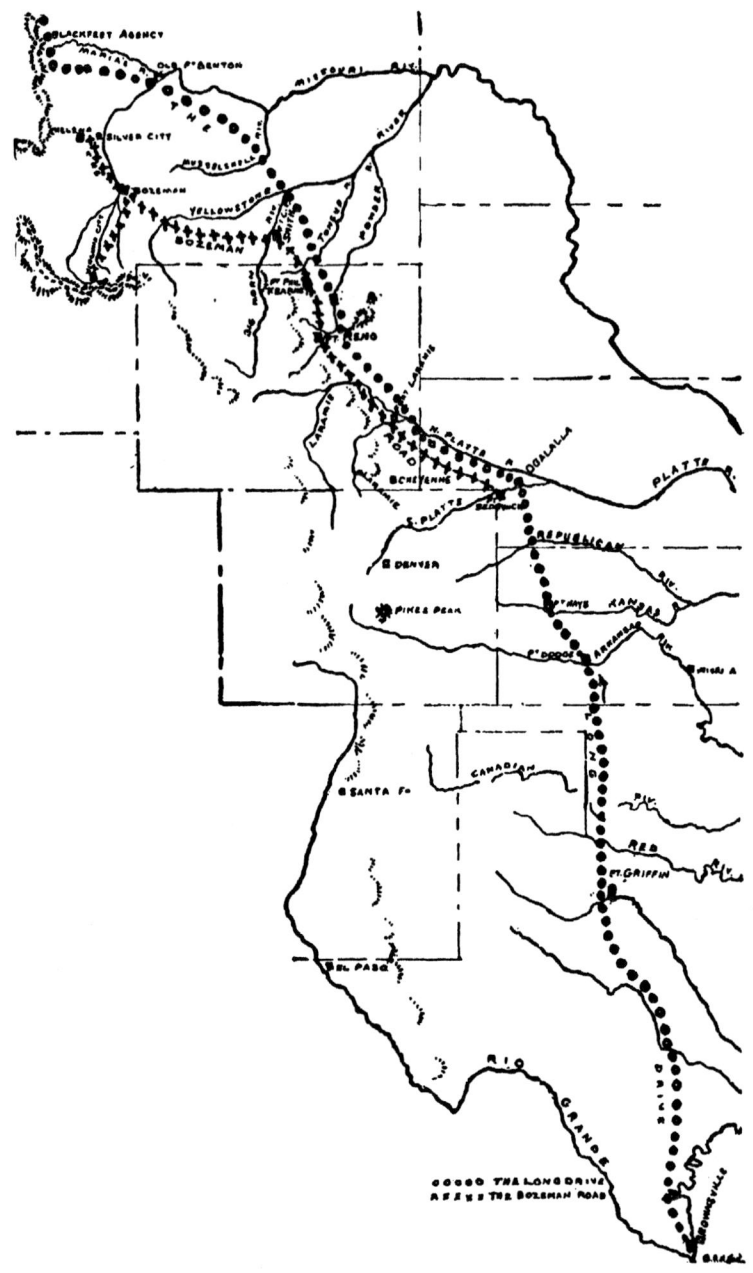

MAP IV. THE LONG DRIVE AND THE BOZEMAN ROAD

○ ○ ○ ○ ○ The Long Drive. × × × × × The Bozeman Road.

Creek, another branch of the Powder River, and still farther north on the Big Horn the government erected Fort C. F. Smith, the first two forts being in Wyoming, the last in Montana.

Brady says: "There never was such a post as Fort Kearney for real trouble. It was in a state of siege for the two years of its existence; war parties hidden in the woods and mountain passes constantly kept their eyes on the post, so that no one was safe outside of the stockade. The Indians attacked wagons and trains, stampeded the cattle, and ran down detached parties from the garrison. If grass was to be mown for the horses, or wood to be cut for the post it all had to be done under a heavy guard, and although the country was full of wild game there was no hunter that dared venture beyond the camp for fear of the silent and hidden red man."[1] The mountain men had learned that when the Indians were not to be seen they were the most dangerous. Bridger always said: "Whar you don't see no Injuns there they's sartin to be thickest."

In order to complete Fort Phil. Kearney it was necessary for the wood-choppers to go to the heavy timber for logs, though not a day passed when the Indians did not appear and attack the little detachment of workers. One day Colonel Fetterman with eighty-one men went out from the fort to protect the wood train. As Fetterman approached the place where the choppers were at work, the Indians attacked the soldiers and then fell back over Lodge Pole Ridge, with Fetterman and his men after them, though the express command had been given by Colonel Carrington, the commandant at the Fort, that if an attack was made no one should cross the ridge.

Immediately after Fetterman disappeared over the fatal

[1] Indian Fights and Fighters. Copyright, 1904. McClure, Phillips & Co.

ridge the fort people heard heavy firing, and Lieutenant Ten Eyck with fifty-four men hurried to the relief. But when the detachment crossed the ridge not one white man was found alive. Eighty-one mutilated bodies, killed by bullets, arrows, hatchets, and spears, lay on the hillside where to-day stands the rough cobblestone monument by the stage road between Sheridan and Buffalo (Wyoming), commemorating the slaughter of Fetterman and his brave men. On the shield of this monument may be traced these words:

ON THIS FIELD ON THE 21ST DAY OF
DECEMBER, 1866,
THREE COMMISSIONED OFFICERS AND
SEVENTY-SIX PRIVATES
OF THE 18TH U. S. INFANTRY AND OF THE
2D U. S. CAVALRY AND FOUR CIVILIANS
UNDER THE COMMAND OF CAPTAIN BREVET-
LIEUTENANT COLONEL WILLIAM FETTERMAN
WERE KILLED BY AN OVERWHELMING
FORCE OF SIOUX UNDER THE COMMAND OF
RED CLOUD.
THERE WERE NO SURVIVORS.

Major Powell was in command of the men who were guarding the wood-cutters on the day of the wagon-box fight. This time the wood-cutters were getting timber to burn and were carrying it to the fort in wagons, the boxes of which were made of strong and thick wood.[1]

When the Sioux came on, three thousand strong, many of them well armed with the army muskets captured during the Fetterman fight, Powell and his men made a corral or fortification out of the wagon boxes. In these boxes the soldiers crowded, and from this vantage-point did their shooting. Not finding room for all to shoot from so small an enclosure, Powell selected the best shots to use the

[1] Some historians claim that these wagons were made with iron bottoms.

rifles, and set the rest of the soldiers to loading the guns as fast as they were emptied. Time and again the Indians charged on the handful of men, only to be repulsed each time with heavy loss. The great chief, Red Cloud, wild with fury

FETTERMAN MASSACRE MONUMENT

that his three thousand warriors could be so driven back by thirty men, led the charge in person. But Indians could not withstand such a fire as the little band poured forth. Even Red Cloud fled. This ended the fight. Powell lost three men; of the Sioux were killed or wounded at least eleven hundred warriors. A sweet revenge for the Fetterman massacre!

After this battle Fort Phil. Kearney was abandoned by direction of the government. The Indians afterwards burned the hated buildings to the ground.

"It was never reoccupied, and to-day is remembered simply because of its association with the first, and with one exception the most notable, of our Indian defeats in the West, and with the most remarkable and overwhelming victory that was won by soldiers over their gallant red foeman on the same ground." [1]

2. THE CHEYENNE UPRISING

The Indians along the line of the Oregon Trail, between Council Bluffs and Fort Laramie, had very serious objections to the construction of the Union Pacific Railroad across their lands. When the Kansas Pacific was being constructed toward Denver, again the Indians arose and warned the white man that if it was railroad it also was war. Chief Roman Nose of the Cheyenne tribe, said: "This is the first time that I have ever shaken the white man's hand in friendship. If the railroad is continued I shall be his enemy forever."

The Indians believed that they had been successful in their fights along the Bozeman Road, notwithstanding their heavy loss, because the government had abandoned the Powder River country, and Forts Reno, Phil. Kearney, and C. F. Smith. When the troops were withdrawn from these forts, the Indians rejoiced in their hearts and firmly believed that our government was afraid of them. This feeling of triumph over the white man gave other tribes encouragement and had much to do with the uprising of the Cheyenne tribe.

[1] Brady. Indian Fights and Fighters.

We must admit that the government was not always without fault in its dealings with the natives. Treaties were easily made and as easily broken by the white man. Certain tracts of land were portioned out for the hunting-grounds of the red man, to be taken back again 'when they seemed of special value. The Indians had an idea that a treaty once made was for all time, and they naturally insisted by force of arms upon the enforcement of the treaty. They also believed that the lands over which their fathers and fathers' fathers had roamed and hunted and trapped were theirs by right of inheritance. This mutual distrust existing between the red man and the white man is aptly illustrated by an incident. A missionary to the Indians on a certain occasion went to preach to one of the tribes. The white man wore a handsome fur coat which he took off just before he began his sermon. "Where shall I put this to keep it safe?" said the bishop to a chief. The warrior at once replied, "Put it right there on the snow, there is not a white man within twenty-four hours' journey."

"In the fall of 1866 the Cheyennes swept through western Kansas like a devastating storm. In one month they cut off, killed, or captured eighty-four different settlers, including women and children. They swept the country bare. Again and again different gangs of builders were wiped out, but the railroad went on. General Sheridan finally took the field in person, but with an inadequate force at his disposal." With Sheridan was a young cavalry officer, Colonel George A. Forsyth, who was eager to take command of some troops and go into the field in an active campaign against the Indians. With the permission of Sheridan he raised a company of forty-nine scouts of the very best type of fighting men. In this company were hunters and trappers and veterans of the Civil War. Forsyth followed the trail of a large hostile band to the banks of the Arickaree and there went into camp, intending to take up the pursuit again in the morning. But in the morning it was the Indians that did the pursuing. They surprised the camp, and, but for Forsyth's alertness, would have captured the entire force. As it was, they stampeded the pack-mules and compelled the little party to retreat to a small island in the river. Here shallow rifle pits were dug with bowie-knives and tin plates, and a warm reception was prepared for Roman Nose, who led on his warriors in person. They were seven hundred strong and held both banks of the river, from which they poured a galling fire upon the devoted fifty. This killed every one of the cavalry horses, whose bodies were used as a rampart, and whose flesh was eaten raw by Forsyth's men during this long fight.

Forsyth himself was shot in the thigh, again in the leg, and still later in the head, but he fought till relief came, like the hero he was. Roman Nose finally determined to end all with a charge which he led in person. He was a great leader, and he led not one but six good charges, probably the best ever made by Indians. Up the almost dry bed of the stream they came, well mounted, yelling like demons, and led by their gigantic chief, who shook aloft a rifle captured in the Fetterman massacre, and swore that he would add to it the rifles of all those opposing him that day. But he never got those rifles, nor did he need them evermore, for after performing wonders of bravery, bringing his undisciplined followers to the very muzzles of the guns again and again, rallying them after each disastrous charge and inspiring them with something of his own courageous spirit, he fell in the sixth charge, literally shot to pieces.

The Indians fell back wailing over the loss of their chief. When the smoke of battle had cleared away it was found that twenty-one of the forty-nine whites could no longer do service. Some were dead, others disabled.

There was nothing to eat except the raw meat of the dead horses which had so long been exposed to the hot sun that their carcasses were in a horrible state. But here the company held its ground for eight days, besieged by the Indians until succor came from Fort Wallace, whither Forsyth had sent two volunteers, under cover of night after the first day's battle. On the sixth day Forsyth assembled his men, and advised those who were well enough to leave the island that night and to endeavor to get to Fort Wallace, while he and the rest of the wounded would run chances of escape. Not one would go. They would live or die with their comrades. Does Thermopylæ's defense deserve eternal commemoration any more than this of Forsyth and his gallant forty-nine on the Arickaree?

The Cheyenne war was ended by Colonel George A. Custer, sometimes called by the Indians "The White Chief with the Yellow Hair," and sometimes "Long Hair." In the dead of winter, with the thermometer at zero and deep snow on the ground, Custer marched with his regiment to the frozen waters of the Washita, surprised in camp the worst band of the Cheyennes, led by Black Kettle, and charged them so fiercely that the few who survived were glad to surrender. After the destruction of this band the others came in, surrendered, and were sent back on the reservation. where they behaved fairly well for six years.

3. THE APACHES

When we acquired title to New Mexico, through the treaty of Guadalupe Hidalgo in 1848, we also fell heir to the Apache tribe of Indians, who roamed in what is now New Mexico and Arizona. The Apache had a cruel face, was brutal and fiendish to an unusual degree. While the Sioux was brutal, he was willing to fight in the open. The Apache was a lurking skulker whom General Crook called "a tiger of the human species."

Ever since the white man had gone into the region of the Apache there had been war and massacre of women and children. In 1871 General George Crook was put in command of this department with orders to tame these fierce aborigines. This officer had won his spurs as an Indian fighter in other fields, and few officers knew nearly so well as he how to handle Indians. He believed that the Apaches should be placed on a reservation and made to stay there and earn their living.

GENERAL GEORGE CROOK

He further believed that temporizing should be abandoned and an energetic campaign waged against the tribe.

Crook was a just and true friend of the Indian, but at the same time was an able and clear-headed officer. His first work was to call the Apache chiefs together and tell them that if they wished to stay on the warpath it meant extermination; that civilization and the white man were coming into the territory over which the native hunted; and that it would be to the advantage of both white man and native to establish peace.

Then Crook offered them protection if they would surrender and settle down on the lands allotted to them by the government. Many of the Indians accepted the terms and peacefully surrendered, but others refused, and, led by Geronimo, carried on for years a war of surprisal, ambuscade, and sudden retreat that taxed Crook's resources to the utmost. It was in this war that General Leonard Wood, now chief of staff of our army, first won distinction. The lamented Lawton, whose gallant service and untimely death in the Philippines are fresh in the memory of us all, was also one of Crook's best officers.

The Apaches were a difficult tribe to subdue, owing to their bravery, treachery, and endurance. General Crook tells about an Apache whom he saw run for fifteen hundred feet up the side of a mountain without any sign of fatigue. A band of these Indians could ambush a party of white men on the open prairie where there was not a blade of grass, a cactus, or a stone behind which to hide. They would burrow down into the sand until their bodies were all covered and there remain motionless until the white man came almost upon them.

They were familiar with the country, "which they knew as well as if they had made it themselves." They knew every ravine, pass, valley, or cañon in New Mexico and Arizona, and could find hidden places inaccessible and unknown to the soldiers. A great difficulty in the way of their capture was that when they were close pressed, they would not all take flight in the same direction, but would scatter to all points of the compass until no two were in the same place.

Crook was called north to fight the Sioux before he had

put the finishing touches to the Apache war. General Miles has the distinction of having ended this long war after he had already ended the Nez Perce and Sioux wars in the north. Miles was one of the greatest Indian fighters we ever had. After a final whipping from Miles, Geronimo came into camp, boasting that it was the fourth time he had surrendered to the white man. Miles saw to it that he never surrendered again, for he sent him and the worst of his band to Florida. Geronimo never saw his beloved mountains again, though he was finally permitted to settle on the little Apache reservation in Indian Territory and died there very recently. Geronimo's wars cost our government over one million dollars, much loss of life and many bloody combats.

GENERAL NELSON A. MILES

After the Apaches were sent away, Arizona and New Mexico drew many settlers to their sunny lands, a thing utterly impossible while the valleys were open to the sudden and fierce raids of the natives, who may with justness be called the most cruel of all western Indians.

4. THE SIOUX

When the year 1876 came our government decided that all of the Indians in the Northwest must stay on reservations that had been selected for them as homes. On the other hand, the Indians decided that they would roam and live where the hunting and fishing were the best. Now, when the government and the Indians did not agree upon any policy there was always trouble, and this case was no exception to the rule. The trouble at this time is known as the Sioux war, because Sitting Bull and his Sioux were most prominent, but the Cheyennes, too, were numerous in the hostile bands. The contested land on which the Indians determined to roam was encircled by forts and agencies. The Missouri River enclosed it on the east and north, on the south were the military posts along the line of the Union Pacific Railroad, to the west were the Rocky Mountains.

Three lines of attack were planned against the Sioux to drive them from the hunting-grounds and to get them into their allotted reservations. One under General Gibbon from western Montana, one led by General Crook from the south, and one under General Terry from Fort Yankton, reinforced by Custer's cavalry from Fort Abraham Lincoln, near Bismarck. Our government believed that any one of these armies would be powerful enough to overcome the rebellious Indians. But certain facts were overlooked; namely, that the Indians now had firearms of the highest grade, were well supplied with ammunition, were operating on interior lines, as the army man would phrase it, and had a thorough knowledge of the country. Besides, they were in great force, numbering anywhere from four to eight thousand. In fact, the government had no suspicion that the uprising had become so formidable.

General Crook had been called from his brilliant work against the Apaches of New Mexico and Arizona to help put down this uprising in the north. With his column, in the spring of 1876, he advanced north by way of the Bozeman Road, past the ruins of Fort Phil. Kearney to Tongue River, near the line between Wyoming and Montana. At the arrival of the soldiers in this territory, the Crows and the Shoshones, hereditary enemies of the Sioux, joined Crook's army, wishing to have a hand in the annihilation of the common foe. · The exact location of the enemy was not known, but the Sioux were supposed to be on the Big Horn, Rosebud, or Powder rivers.

While Crook was going north, Terry with Custer moved up the Yellowstone to the present site of Miles City, and waited for the division under Gibbon to come from western Montana. Crook was the first to encounter the Sioux, who were led by Crazy Horse. It was in the valley of the Rosebud and the fight was long and fierce. Never had Crook seen Indians fight so desperately. In fact, the natives came very near making it a bloody defeat for this old Indian fighter. It was afterward learned that Crazy Horse had planned an ambush for Crook and had almost succeeded in his cunning design. Crook was finally obliged to retreat and wait for the other officers to appear upon the field.

Now, Custer had not heard of Crook's attack and retreat when he found the enemy on the Little Big Horn, under the leadership of Gall, Crazy Horse and Sitting Bull. When Custer dashed into the camp of the Indians, he did not know the strength of the Indians' force, which in reality was of four to six thousand warriors; nor of the excellent firearms which they possessed. Moreover, Sitting Bull was there to plan the fight, while Crazy Horse and Gall were great leaders in battle. Custer divided his six hundred men into four unequal divisions, placing Major Reno in the command of one, with instructions to attack the village at its lower end, while Custer should enter it farther up. Reno's attack was a failure, though Custer did not know this and expected the support of Reno and his men when he made the attack.

With two hundred sixty-two men of the Seventh Cavalry Custer dashed upon the Indians. But he never came back, neither he nor any of his gallant band.

GENERAL GEORGE A. CUSTER

With Custer fell his brothers, Captain Tom Custer and Boston Custer, Calhoun, his brother-in-law, and Autie Reed, his nephew. Michigan has erected a bronze monument in Monroe, the home town of the Custers, to commemorate this day, fatal for the Custer family.

The Custer fight practically ended the summer's campaign

with the Sioux. While our soldiers were humiliated by their defeat, the Indians were amazingly puffed up. Now, they believed that they could whip the "Big Knives" and drive the last white man off their lands. Those Indians who before were willing to remain on the reservation in quiet and peace now stole away to join the successful warriors; and the problem then confronting our government was how to keep any of them on the reservation.

General Wesley Merritt with the Fifth Cavalry was on his way to reinforce Crook when he received word that some thousands of Cheyennes had left the reservation at the Red Cloud Agency and were hurrying to join the triumphant Indians. Merritt had for his chief of scouts Buffalo Bill (William Cody), who knew the country as well as any of the Indians. His first move was toward the north to head off the Indians at War Bonnet Creek. Buffalo Bill was riding ahead with fifteen scouts when he met the advance guard of the Indians. Yellow Hand, their chief, rode out ahead and challenged Bill to a duel, saying: "I know you Pa-he-haska (Long Hair). If you want to fight come and fight me."

At full tilt these two warriors galloped toward each other and fired. Yellow Hand was shot in the leg and his horse killed, but at the same time Buffalo Bill's horse fell. Only twenty paces apart they fired again. The Indian fell, the white man was unhurt. Buffalo Bill, snatching the war-bonnet from the head of the dead chief, and waving it above his head yelled, "The first scalp for Custer." This was the only blood shed in that raid. The Cheyennes submitted to be driven back on the reservation, and Merritt went on and joined Crook in his camp on Big Goose Creek, near the present Sheridan, Wyoming.

During the winter General Crook defeated American Horse at Slim Buttes, and forced all of the bands under his command to surrender. Then General Miles built a cantonment opposite the mouth of Tongue River, near the present site of Miles City, Montana, and executed a well-planned campaign. In the dead of winter, with snow deep on the ground and the thermometer at zero, and sometimes away below, Miles raided the camps of the Indians, who had their quarters in the protected ravines of the Rosebud, Tongue, and Powder rivers. Crazy Horse made a stubborn fight but could not hold out against this unremitting warfare. One by one the bands came in and surrendered, until all save Sitting Bull's were back on the reservations. Sitting Bull fled to Canada, where he found safety. Later he returned to the United States, where, after living a peaceful life for a few years, he was killed by an Indian policeman who was attempting to arrest the old chief by orders of one of our army officers. This was the last real war with the Sioux, though in the early nineties they made serious trouble at the Standing Rock and Wounded Knee Agencies in South Dakota, and for a time people in the West dreaded a return of the Sioux raids of 1862, when they killed thousands of settlers in southern Minnesota and northern Iowa; or the still worse Sioux war of 1876, when they fought our best western troops to a standstill.

5. CHIEF JOSEPH

The Nez Perce Indians had been at peace for many years. They were a peaceful tribe and made the boast that no white man's scalp had ever hung in their wigwams. This tribe had been given a home on the Lapwai reservation, just east of Lewiston, Idaho, which had been disdainfully declined by their chief, Joseph, who was determined to have this tribe return to the Wallowa Valley where their people had lived for generations. Joseph had a strong love for the soil of his hereditary hunting-ground. Once in council he said, "A man who would not love the ground of his father and mother is worse than a beast."

When in the summer of 1877 it became plain to Chief Joseph that he must go upon the reservation or fight, he chose the latter alternative. In two well planned battles he whipped the forces of General Howard, and then gathering all the Nez Perces that would follow him, men, women and children, he beat a masterly retreat. Eastward through Idaho he went, beating off the troops that hung on his skirts and killing settlers here and there. Encumbered as he was with numbers of non-combatants and much baggage, yet he won through Lolo Pass without loss, thus crossing the Bitter Root range at the very point made famous by the memory of Lewis and Clark. He emerged into the Bitter Root Valley a short distance south of the present pretty town of Missoula and here turned south. This was his great mistake. Had he turned north, through Hell Gate Cañon, the war trail of the Flatheads and Blackfeet, nothing could have hindered his escape to Canada, where Sitting Bull had found a refuge the previous year. It may be that he was not yet ready to admit that he must expatriate himself so completely from the land of his fathers. It cannot be that he was afraid of the feeble garrison at Fort Owen in this same Bitter Root Valley, for he brushed contemptuously by them in his flight to the south. General Gibbon gathered a force, pursued him, and pounced upon him in the valley of the Big Hole, ninety miles northwest of Dillon, Montana. But Joseph was no ordinary chief and his were no ordinary Indians. They hurled back Gibbon as they had hurled back Howard and got clear away with smaller loss than their adversaries sustained. Now, with both Howard and Gibbon on his rear, Joseph went over the mountains again into Idaho, eastward into Wyoming, through Yellowstone

Park, frightening away the tourists, who even in those early days were there in surprisingly large numbers; and then, seeming to resign himself to the inevitable, he turned north and began a race through Montana in the hope to reach Canada; and for all Gibbon or Howard could have done he would

WHERE THEY HAVE SNOW IN AUGUST.   SYLVAN PASS IN
YELLOWSTONE PARK

have reached it.   But General Miles with his seasoned Indian fighters was away east on the Yellowstone, lying in cantonment at Miles City.   To him came an express from Colonel Sturgis, who had been closest on the trail of Joseph all the way through Yellowstone Park.   The great Indian fighter set out at once for the Missouri, to cut off the great Indian leader.   It was a fierce race, but the white man won.   At Bear Paw Mountain, scarcely two days' march from the

Canada line, he forced Joseph to give battle, and a real battle it was. The Indians stood at bay so fiercely that Miles failed in several attempts to storm their position. He was in serious difficulty, for any moment might bring Sitting Bull and his bands, who were not far away in Canada. But the forces of Hunger, Cold, and Fatigue fought on the side of the white man. On the fourth day Joseph surrendered, saying: "It is cold and we have no blankets. The little children are freezing to death. My people, some of them, have run away to the hills and have no blankets, no food. No one knows where they are, — perhaps freezing to death. I want to have time to look for my children and to see how many of them I can find. Maybe I shall find them among the dead. Hear me, my Chiefs! My heart is sick and sad. From where the sun now stands I will fight no more forever." And he never did.

Poor old Joseph. They sent him up into Washington on the Fort Colville reservation. Up to very recent years the old man could be seen there brooding over the fire and saying no word to anyone. Who can say what mournful pictures he saw in the flames? The lands of his fathers gone, his braves sacrificed for naught, his wonderful retreat of no avail, his people scattered and impoverished. Yet in spite of his failure, Chief Joseph deserves a lasting place in history. No one can deny him the title, "The Indian Xenophon."

6. THE UTAH INDIANS

The Mormons believed that it was cheaper to feed the Indians than to fight them. Carrying out this policy, they gave the Indians lessons in agriculture, taught them to plow the land and to raise crops. Notwithstanding all of this, the Utes again and again broke out in open hostility, and to this day you can see in Salt Lake City, Ephraim, Manti, and

many of the towns remains of the old rubble walls that were
built as fortifications.

It was during the year 1853–54 that one of the fiercest
conflicts raged, at which time many lives were lost, many
inhuman acts committed, and a great deal of property was
destroyed. This war was called the "Walker War" because
it was led by Chief Walker. This chief was a great favorite
with the Indians because he was athletic, a prime shot, and
could speak not only many native languages, but Spanish
and English. "When he used to go forth to battle, he dressed
in a suit of finest broadcloth cut in the latest fashion, and
donned a cambric shirt and a beaver hat. Over this costume
he wore his gaudy Indian trappings."

When the Mormons first came to his country Chief
Walker gave them a hand of welcome and helped them select
many choice pieces of land. But when he saw his choice
hunting-grounds turned into farms, and the game being
rapidly driven from the lands, he raised his hand against
the white man and attempted to drive him from the country.
After frequent massacres of the passing emigrants and
settlers a treaty was made establishing temporary peace.

Many Indian outbreaks occurred between 1857 and 1862,
no real curb being placed on the natives until General Conner
with his volunteers, early in 1863, successfully fought the
Shoshones and Bannocks, led by Big Hunter, Pocatello, and
Sanptich. These tribes had killed and robbed the emigrants
and plundered the overland mail route until conditions
became unbearable. In the battle of Bear River, though the
cold was intense and the snow was so deep that he could
not use his cannon, Conner almost annihilated the Indians.
This defeat checked their fighting spirit and the good results
of the battle were immediately felt, particularly throughout
northern Utah, where cattle and horses were now safe and
settlements could be made without fear.

For two years, 1865–67, Chief Blackhawk waged an ugly war against the settlers in southern Utah. So widespread became the alarm that many of the settlers abandoned their farms and homes and left the district in possession of the natives. The volunteers and militia, who served for two years without pay, did valiant service, and pensions are still granted to those Blackhawk veterans.

The Shoshones and Bannocks finally agreed to a treaty in 1863, whereby they received an annuity for a term of twenty years. The Utes also entered into an agreement with the government in which they relinquished all claim to their lands and agreed to live on the reservation, where they received large money annuities, dwelling houses, cattle, farms, and schools. Chief Blackhawk went with his tribe and settled down to the peaceful duties of a farmer.

7. THE MODOCS

The most costly war that the United States ever had, considering the small numbers engaged in it, was the Modoc war of 1872. The Modocs are a branch of the Klamath tribe, who killed Smith's men and Fremont's men, and when the Modoc war began they were living near Klamath Lake. This was a beautiful country and it was agony for the Modocs to give it up and remove to the reservation picked out for them by the government; but they did it, and seemed resolved to make the best of a hard case. But the reservation was too small. The Indians already there, Klamaths chiefly, resented the intrusion, taunting them with the epithet "outcasts," and telling them that they were too poor to have a reservation of their own, so they had to live on the land belonging to another tribe. This angered the Modocs and they deliberately went back to their former hunting-grounds near Klamath Lake. By this time the settlers had commenced to come into that region, and were building their homes and cultivating their land. Barbarous assaults were reported to the government, and petitions were sent to the Indian agents to have the hostile natives driven back to the reservation.

At this time Brigadier-General Canby, stationed at Portland, had charge of all the Indians in that region. Now, Canby was a most fearless soldier, but he was also a just man and he saw that the Modocs had a cause for hostility. He tried to have a special reservation set aside for them, but could not accomplish it, and received orders to make the Indians go to the Klamath reservation. He was ordered to remove them, "Peacefully if you can, forcibly if you must."

A command was then issued for the arrest of Captain Jack, Black Jim, and Scarfaced Charley, who were camped with about fifty warriors on Lost River, not far from Tule Lake. The troops completely surprised Captain Jack and his braves, who immediately opened fire, which was rapidly returned. After a few were killed on both sides the Indians, with their women and children, fled to the lava-beds south of Tule Lake.

These lava-beds had been the natural roaming grounds of the Modocs, and they knew them better than any one else. Indeed, no white man knew anything about them. The lava-beds were of a most peculiar formation, due to volcanic action. When the soldiers looked over the mass, which was eight miles long and about four miles wide, the entire surface looked like a level stretch of land covered with sagebrush; but upon a close examination the ground was found to be much broken, with rocky ridges, and many caves and hollows from ten to twenty feet high. These occurred in groups with parallel lines, re-entrant angles, natural bastions, and enfilading trenches, making a fortification, built wholly by nature, more ingenious and effective than any ever devised by the best military engineers. These ridges were split open at the top, leaving a space five to eight feet wide in which a man could walk or crawl from rampart to rampart without being seen. No wonder that Captain Jack boasted that his little band could stand off a thousand soldiers. He and his men knew every rock, the soldiers knew nothing of the country.

Into this natural fortress, Captain Jack took his eighty warriors, ammunition, food, women, and children. Once there with abundance of good water, he was able to beat off any ordinary attack, and the region was too extensive to permit of any effective siege with the small force at General Wheaton's disposal. General Frank Wheaton, an officer of much experience, had several troops of cavalry and infantry, two companies of Oregon volunteers and one of the California Volunteer Riflemen. All of one day soldiers walked over the lava-beds, scarcely seeing an Indian. "They stumbled blindly forward over rocks ranging in size from a cobblestone to a church with points like needles and edges like razors." Suddenly smoke would be seen from the side of some rock and a soldier would fall; but a return volley only hit the rocks.

The situation was reported to Canby, who still insisted that peace should be made and the Indians given a new reservation. He now took the field in person, but before proceeding to extremities tried once more what could be done through negotiating. A peace conference was then had through the interpreter Riddle, who had with him a Modoc squaw wife. This squaw, Tobe, had a kindly feeling for the white men and warned them against the treachery of the Modocs; but Boston Charley and Bogus Charley had come to General Canby saying that Captain Jack wanted to surrender and wished to make terms before a peace commission. Jack sent word that he and five of his warriors, all unarmed, would meet the white men in a tent set up half way between the opposing forces. Canby and four men with Riddle and Tobe went to the council-tent, though the squaw all the time protested and declared that they would all be murdered. Soon after the conference began it was plain that treachery was in the hearts of the Indians, but relying upon coolness and audacity to carry them through, counting rather too much upon the trained power of the white man to awe the wild red man, Canby spoke with as much authority as if he had the whole United States army there. Suddenly at the signal, "At-tux" (All ready), the Indians drew their concealed revolvers and fired, then fled. General Canby and two of the men were killed, one of the men with Riddle escaping. The squaw was knocked down, but was finally rescued by one of her own tribe.

Additional troops were soon sent over the lava-beds, but the enemy had fled, though at one time Captain Jack was seen wearing Canby's uniform. Out in the open, they were finally run to earth by the cavalry, so tired and weary that Jack said, when they found him sitting on a log, "My legs have given out."

In a military trial Captain Jack and four of his braves were found guilty of murder and were hanged.

The Modoc war cost the government half a million of dollars, the lives of one hundred and sixty-eight white soldiers and one of the best and bravest officers that ever graced our army. Yet it must be remembered that Captain Jack was a chief who was firmly opposed to war, and that at the beginning of the Modoc conflict the Indians were in the right and were driven by bad faith and injustice to desperation, with which one might sympathize.

REFERENCES

Carrington. Army Life on the Plains.
Forsyth. The Story of the Soldier.
Hansen. The Conquest of the Missouri.
Brady. Indian Fights and Fighters.
Finerty. War Path and Bivouac.
Miles. Personal Recollections.
Brady. Northwestern Fights and Fighters.
Cody. True Tales of the Plains.
Custer. Life on the Plains.
Wood. Lives of Famous Chiefs.
Bancroft. History of Utah.
McLaughlin. My Friend the Indian.
Whitney. History of Utah.
Paxson. The Last American Frontier.
Boyles. The Spirit Trail.
McBeth. The Nez Perce Indians since Lewis and Clark.

CHAPTER VIII

COWS AND COWBOYS

1. The Long Drive   2. The Cowboy

1. THE LONG DRIVE

A new industry came to the Great West coincident with the coming of the settler. After the military men had subdued the Indians, and the government had placed them on reservations, there were three classes demanding certain kinds of provision: the home-seeker, the wards of the government, and the troops whose duty it was to see that the laws prohibiting the Indian from roaming where he willed were properly executed. Naturally the food most desired was meat.

In the early days the buffalo furnished meat in plenty, but the wasteful habits of the white man and the professional skin-hunter soon led to its complete extermination. The Indian killed only what buffalo he needed for food and clothing, for he knew how closely his own welfare depended upon the preservation of this great game animal. Did not his sinews for bow-strings, war-shields, bed and bedding, wigwam and tepee, saddle and bridle, and his most nutritious food come from the buffalo? It was those who killed for the valuable skin, and those who destroyed for the love of the sport that drove the buffalo from the Indian's hunting-ground, and caused his extermination by wanton slaughter. The hunter and traveler who shot for sport only used the tongues and choice rumps, leaving the rest of the animal to rot on the plains.

No exact estimate can be made of the actual number

209

of buffalo that roamed over the broad plains. This much is known, that during the years between 1868 and 1881 there was expended in the state of Kansas alone the sum of two and one half million dollars for buffalo bones that had been gathered from the prairies. These bones were utilized by the various carbon works of the country, and represented at least thirty one million buffalo.[1]

The number of animals that moved in one great company was actually countless. In 1868 Sheridan and his officers rode for three days through one continuous herd, and in 1869 a Kansas Pacific train was delayed from nine o'clock in the morning until five o'clock in the evening in order to allow buffalo to cross the track. We read of one herd that covered an area seventy by thirty miles.

As the supply of buffalo meat decreased, herds of tame cattle appeared to supply the deficiency. The extreme Southwest around the western coast of the Gulf of Mexico has had wild cattle ranging on the native grass since the "mind of man runneth not." Tradition does not hesitate to attribute the presence of "cows" in this southern region to the coming of Cortez with shiploads of men, provision, and cattle. However that may be, we find the original home of the great ranchmen in the lands 'of southern Texas and eastern Mexico. From this locality cattle were brought to the Northwest as far as Montana, Nevada, and even to the most northern boundary of the United States.

This movement began doubtless for the same reasons as the annual migration of the buffalo occurred. Spring comes early in the sunny southland. Here the fresh green grass first appeared, and here grazing began earliest. Just as the buffalo slowly ate his way northward all summer, moving constantly toward the land where the summer days were cooler and the herbage was fresher, constantly away from

[1] Inman.

the land where the fierce summer sun scorched the grass and dried up the water holes, so the domestic herds, guided by their cowboy attendants, grazed northward throughout the summer. Fall found them far north in Wyoming or Montana, where they would be sold to stock the ranches springing up there, or to supply meat to the thriving towns of the northland. This was the "Long Drive," one of the most picturesque features of life in the Great West of a generation ago.

After going over the tableland of western Texas, the trail continued north through the Indian country into Kansas, crossing the old Santa Fé Trail at Dodge. When this point was reached the cowman felt that the first half of the long journey was completed. Here many of the cattle, sometimes an entire herd, would be sold after the Kansas & Pacific, building west, came to furnish quick transportation to eastern markets. As the railroads pushed gradually westward new towns sprang up and became emporiums for the cattle trade. Thus Abilene, Ellsworth, Wichita, Dodge City, Hays City, Ogalalla, and Cheyenne became in turn notorious. At the western terminus of the railroad the trail of the cowboy cut the trail of the railroad man. It was Tartar meeting Hun. The cowboy had money to spend, for his herd had been sold and he had been paid off. The railroad gang had frequent pay days. Both were thirsty souls and red liquor was handy. The cowboy was more expert with the revolver, but the construction crews in those days were generally Irish, and terrible men with their fists. Besides, they could generally reckon on the support of the gamblers and roughs, who followed the railroad, and who could pull a gun with celerity and use it with deadly precision. In most of those towns, which a generation ago were on the frontier, the oldest graves are those of men who "died with their boots on."

After crossing the headwaters of the Solomon up past Fort Hays and over the Republican, the next stop was on the South Platte where, not far west of its junction with the main river, was a large cow camp, Ogalalla, the rendezvous of the cowboys and the Texas rangers. This was a typical frontier town,—outfitting place for the cowman, a home of the border ruffian, and a Monte Carlo for the gambler. If there were fifty houses in the town, forty of them would be saloons, gambling dens, and dance-halls. This frontier place was a town of no night, for after the sun went down the harvest from all kinds of lawlessness and crime was reaped. These were the times that tried the souls of law-abiding citizens. Some thrilling instances of daring in defense of order are embalmed in the records of these towns; and foremost among the names of many brave and capable police officers stands that of James B. Hickock, better known as Wild Bill. Now, Wild Bill was a marksman of note even for the frontier, and no man could pull a gun quicker. At the first election in Hays City, Kansas, he was chosen marshal. No desperado that disputed his authority lived to repent it. He was the terror of evil-doers. Members of the McCandlass gang once leagued together to put him out of the way when he was a station guard on the mail route, and at one time a roomful attacked Wild Bill alone. When the smoke had cleared away it was found that ten men had been killed, and Wild Bill had received three bullets, several buckshot, and numerous knife cuts. It took him six months to recover, but after this he was reputed to bear a charmed life. He, too, died with his boots on, but the shot that killed him came from behind. He is buried in the cemetery at Deadwood, South Dakota, and a simple shaft of fine workmanship marks his final resting-place.

Not all of the cattle were sold at the railroad towns. Many went on to stock the ranches of Wyoming and Montana. From Ogalalla the drive was along the Platte over the Oregon Trail to Fort Laramie. Many long drives had to be made in Wyoming without water, and part of this drive was over the Bozeman Road. Part of the time the trail skirted the Black Hills, and again went to the west into the foothills of the Big Horn Mountains. The headwaters of the Powder and the Tongue, the hunting-grounds of the Crows and Sioux, the home of the trappers and the scene of many a conflict with the Indians, were now marked by the trail of the cow. Thus came the newer civilization, the next step in a great western development.

When Frenchman's Ford was reached, at the junction of the Big Horn with the Yellowstone, where trappers' post and soldiers' fort had held sway each in its turn, another fringe of civilization was reached, — the first since leaving Ogalalla. Here were to be found the Indian, the gambler, the trader, the cowboy, and the railroad surveyor, for the Northern Pacific was about to tap this wilderness as the Kansas Pacific and the Union Pacific had done farther south. The most marked difference between these frontier men of the North and those of the South was that in place of wearing one six-shooter at their belts they wore two!

After trailing along the tributaries of the Yellowstone to the north, the rest of the Long Drive was north and west until the Missouri was reached, when the Lewis and Clark trail was followed to Maria's River. The country of the Blackfeet was again invaded; this time in a peaceful manner because the tribe had been subdued and placed on a reservation. The herd now entered the old hunting-grounds where the buffalo had roamed and the beaver had made his dam. It took five months to come from the plains of Texas to the ranges of Montana over the open prairie and through the foothills. Sometimes the cattle had different destinations, for Wyoming, Nebraska, and Idaho received their share of

the southern stock. This easy, definite movement toward
the north was not an experiment: it was a permanent step
toward the occupation of those regions where only a few
years before the buffalo and antelope had eaten of the
nutritious grass, and quenched their thirst from the streams

THE COWBOY AND COWPONY

of pure, sweet water. In the year of 1871 over a half-million
cattle were brought north over the Long Drive.

"It was a strong tremendous movement, this migration
of the cowmen and their herds, undoubtedly the greatest
pastoral movement in the history of the world. It came with
a rush and a surge, and in ten years it had subsided. That
decade was an epoch in the West; the City of Cibola began.
The strong men of the plains met and clashed and warred
and united and pushed on. What a decade that was! What
must have been the men who made it what it was! It was an

iron country, and upon it came men of iron,—dauntless, indomitable. Each time they took a herd north they saw enough of life to fill in vivid pages far more than a single book. They met the ruffians and robbers of the Missouri border, and overcame them. They met the Indians, who sought to extort toll from them, fought and beat them. Worse than all these, they met the desert and the flood, and overcame them also. Worse yet than these, they met the repelling forces of an entire climatic change, the silent enemies of other latitudes. These, too, they overcame. The kings of the range divided the kingdom of free grass."[1]

2. THE COWBOY

At one time in our history New Mexico, part of Arizona, Colorado, Wyoming, Montana, and the western part of Texas, Kansas, Nebraska, and the Dakotas had their wide areas devoted to stock-raising, and this was the only industry carried on in those regions, except in the mountains where there was more or less of mining. During this period of our development the cattlemen and their cowboys were the barons of the plains, and ruled over their domains with the branding-iron and the lariat.

The cowmen, although distinctly a product of the South, in many cases did not return to the original home of the cattle, but stayed in the North and made their homes upon the prairies. This cowman was the connecting link between the frontiersman and the settler. The fence has placed the cowboy in past history. He reigns no more. The buffalo has been driven to the zoölogical garden, the Indian to the reservation, and the cowboy to the "Wild West Shows" and "bucking bronco contests." The open range has given way

[1] Hough. The Story of the Cowboy. Copyright, 1897. D. Appleton & Co.

to the irrigated farm, and the cowboy is speedily being obliged to abandon his free, happy, and independent life to be placed in history with the explorer, the trader, the trapper, and the early pioneer. But you cannot make a farmer of the cowboy. He has tried it without success. He is as uneasy in a new environment as the English cabby is with the gasoline automobile.

A BUCKING BRONCO. "SILVER CITY" MASTERED

When the cattle business was in its prime, the stock lived on the open plains, summer and winter, without any food but the grass as it grew on the ground; without shelter except from the hills and short sagebrush. No attempt was made to feed or shelter the animals during the days of snow and killing winds. The cows "rustled" for themselves, usually weathering through the cold, short days without much loss of life. But when an unusually severe winter came to the plains, the loss of live-stock was very heavy, hundreds and even thousands of the cattle perishing during a three days' blizzard. Gradually the stockmen realized that it was dollars in their pocket to protect and feed this stock properly during a part of the season. Coincident with this realization came the irrigating ditch and its water to turn the prairies into hay-fields. To-day the meadow is as indispensable to the stockman as the horse was to the cowboy.

The dugout, the hut, the shack, the cabin, the sod house, and the ranch have all been the home of the cowboy, "his special geographical location determining the architecture of his dwelling-place." Yet the latch-string of these homes was always hanging out for the stranger, who found a hearty welcome, and plenty of beans, coffee, bacon, and hot bread, for which no pay was ever expected or accepted.

Every class of men was represented by the "Cow-puncher." It is a very mistaken idea that to be a cowboy one had to be a border ruffian. It is true that many a man deliberately drifted west to be forgotten or to forget, for the etiquette of the plains forbade the asking of one's past history. The cowboy was accepted for what he was, and if he did not fit into his new calling, he soon found it out and sought other fields.

Clothes did not make the man, neither were they his undoing. Young men from the best families of the East came to the plains, seeking the freedom of a western life, to learn the business in which their fathers or relatives had invested their capital. Graduates from our best colleges lost their identity on the democratic prairies, and became cowboys among the cowboys.

When fashion decreed in 1836, a really great date in our western history, that the beaver hat must be discarded for another style, it hit a killing blow at the calling of the trapper and the hunter. The cowboy has never suffered from the change of fashion or style of dress. As he was in the South in his early days so he was yesterday. There was no part of the cowboy's dress that had not been adopted with thought of utility. He wore what he did because it best served his life and his purpose. His shirt was always open at the neck, around which he wore a bright silk handkerchief, carefully knotted, to keep off the sun and keep out the dust — the more brilliant the neckwear the prouder the possessor. It would be difficult to describe his coat, because he seldom wore one except in winter, and then not always. His vest he never wore buttoned, as he had a theory that he would take cold if he did. The cowboys "chaps," a word cut down from the Spanish *chaparejosare* might easily be called the distinctive part of his dress. These "overalls" are two wide trouser legs fastened to a leather and much buckled belt. They are made of leather and serve to protect the legs of the rider when he goes through underbrush and from the winds and storms of both winter and summer. Most fantastic are some of the chaps, made from the skin of the black bear or the beautiful angora goat, while others are of plain leather with a fringe of the same material running from the hip to the foot. These garments are loose, flapping things that give perfect freedom to the cowboy on his bronco.

But there was one part of the cowboy's dress that was always tight, and here he showed his vanity. These were his high-topped boots with their narrow high heels and thin soles. His toes were so cramped that he did not walk naturally but with a hobbling gait, but then the cowboy did not pride himself on his walk, — on the contrary, he was proud that he did not walk. The cowboy would rather have searched for a horse and taken the trouble to saddle it than walk a fraction of a mile. He rode anything, but he never walked. His high-heeled boots were evolved from long experience. The high heels prevented the feet from slipping through the stirrup, and helped to hold the rider firmly in the saddle by the firm grip they took when the foot was thrust solidly forward in the stirrup. So, after all, it may not have been vanity but a sense of fitness that made the cowboy adopt the high-heeled boots.

The cowboy never rode without his fringed gauntlet gloves. These were worn to protect the hands from being burned by the rope, which he used so constantly and with such dexterity.

When we consider the headgear, or "bonnet," of the cowboy we have an article of dress that deserves more than passing notice, for so practical and serviceable has become this piece of wearing apparel that it has been universally adopted by men of the West. The "cowboy hat" is not a cheap, flimsy affair, but is made by the best hatters, and to possess "a genuine Stetson" is the chief delight of the cowboy. These wide-rimmed, light-colored felt hats must serve for all kinds of weather. Wind, rain, snow, and much wearing do not change their shape. They keep the head cool in summer and warm in winter. In fact they are the umbrella in the rain, the shade in the sun.

Tied back of the saddle was the yellow oilskin raincoat, the "slicker of the cowboy." Through this the most driving rain or snow could not penetrate.

It is not difficult to understand why the cowboy adopted a costume so well adapted to his needs. It did not essentially change through the years that he adorned the range, because nothing else was devised that better met his needs or served his purposes.

The bridle, quirt, lariat, and saddle of the cowboy all show a wonderful skill in workmanship. They are made to last, and to do long hard service. The easy, broad saddle not only makes it possible for the cattlemen to remain unlimited hours on their ponies, but its great strength is a necessity in roping cattle, when one end of the rope is around the steer's leg and the other end around the horn of the saddle.

The revolver, once used for self-defense, came to be employed chiefly to turn the herd when in a stampede, or to kill a wounded or disabled animal or to frighten cattle from the heavy brush. Although the cowboy of late years did not have much use for a revolver he became very expert and quick with firearms.

The Spaniards brought not only their cows with them from the sunny South, but also much of their language, which the people of the plains have unconsciously adopted. From these men of the Southwest we have obtained not only the word "chaps," but "bronco" (rough or wild); "quirt" (cuerda, a cord); "lariat" (lareata, a rope); "ranch" (rancho); "sombrero"; "pinto horse" (painted or mottled); "corral" (korral, a yard); "lasso" (lazo, a slip knot); "cinch" (cincha, to gird, belt or belly-band); "taps" (tapadazo, hood, the leather covering to foot in stirrup); "latigo" (cinch strap), the long cinch straps on left side of saddle to tighten and loosen cinch; "hacamore" (bridle or halter); "roundup" or "rodeo," and many other words that are commonly used without a knowledge of their derivation. The language of the plains was racy and picturesque as the possessors of it.

Without fences and no definite boundary to the "free grass," it was impossible for the cattlemen to indentify their stock unless marked with some individual device. The barons of other countries marked their possessions with crests and heraldic signs; the barons of the plains used the branding-iron with its combination of letters, bars, circles, and squares. Originally each cattleman branded his stock with his initials, and if two men happened to have the same initials the distinguishing mark would be a bar or circle combined with letters. There were no cattle of the plains without these individual markings. While it may seem cruel to place a red-hot iron against the hide of an animal until the hair is burned away and an impress stamped on the skin, yet no better way has been devised by which stockmen can identify their animals. As the cattle industry grew, laws regulating brands and branding were made by the legislatures of the different states and territories, and stock associations were formed to regulate and protect the interests of the cattlemen. Finally books were kept by the stock associations, wherein were recorded the brands that the different stockmen used, and no two brands too much alike were allowed.

In the earliest days of the cattle industry it was customary for a cowman to put his special brand on the side or shoulder of any wild animal found by him upon the open range. When the days of the "roundup" came all unbranded cattle found within a definite boundary on the plains belonged to the party having the "roundup." The "roundup" was exactly what the word indicated. When the cattle grazed all winter, going where they willed without guide or protector, they would wander from the selected range of the owner. In the spring the cowboys were sent in every direction from the ranch to collect the stray cattle and drive them into closer quarters. If an animal was found bearing another brand than that owned by the cattle company working the "roundup," the cowboys would cut it out of the herd and have it returned to its rightful owner. If cattle had no brand, they were branded at once and henceforth belonged to the "outfit." When a calf was found following its mother it was always properly branded with the marks that its parent bore. If the little fellow wandered around in an orphan state, it was called a "maverick" and was sold at public auction, the proceeds being divided among the stockmen of that district or going into the treasury of the association.

The honest cattleman had no title to a maverick. The "rustler" branded it as his own. Cattle thieves also changed the brands on cattle belonging to others so that they could claim them as their own. Thus a U F can be easily changed, or burned over to read O E; or ⸗ V can be made to look like Z W by the addition of curves and lines; and ⱴ (lazy K) made to appear as w̲. When the cattle are sold by one stockman to another the brand must be changed in order that the last possessor may be able to identify his cattle and show his authority to possess stock with some other brand than his own upon their hides. In the language of the plains the brands are "vented" by placing an additional brand, or mark, near or through the old brand, which in no case must be obliterated. To completely destroy the old brand would make the title to the cow questionable; to "vent the brand" would show an abstract to title. A brand H̅ I, bar H I, would be vented by placing a bar below the H, or by a crossbar through the brand, H̅ I, or H̶-I̶.

The law of the plains was a just one. The changer of brands, or rustler of mavericks, was figuratively branded "thief" and there was no place for him among the cowboys. He was told to leave the country, and woe to him who disregarded the warning. But there came a day when the rustlers became bolder and banded together like the pirates of old, systematically taking cattle that belonged to others. Then the stockmen had to join forces against these robbers of the plains and a war of extermination began, which in the long run always resulted disastrously to the rustler.

Most ranches were known by their brands. The brand of a stockman would give his ranch and his foreman that name. The M‐ (M bar) made the ranch the "Embar," and the foreman of the company was not Billie Jones but "Foreman of the Embar"; or the concern would be known as the "Two-bar (=) outfit," or ▢‐4 "Mallet Four." There are many post-offices in the Western States that have these brand names, and letters are sent to Two Bar, Wyoming, Embar, and Two Dot, Montana. Sometimes this custom produces combinations very ludicrous to the ears of tender-

ROPING CATTLE ON THE PLAINS

feet, as in the story where "Bar Y Harry married the Seven Open A Girl."

Thus the cowboy has been a factor in the building of the West, and as an empire-builder he deserves a place in history. It has been aptly said, "The cowboy did not make two blades of grass grow where one grew before, but he caused double the number of cattle to graze upon the unutilized grass which made one of the resources of the land. He was not only a product of the country, but a producer for the country, and he distinctly added to the total of the crude natural wealth quite as much as the farmer who digs such wealth up out of the soil." [1]

[1] Hough. The Story of the Cowboy. Copyright, 1897. D. Appleton & Co.

REFERENCES

Hough.　The Story of the Cowboy.
Adams.　The Log of a Cowboy.
Roosevelt.　Ranch Life and the Hunting Traii.
Wister.　The Virginian.
Talbot.　My People of the Plains.
Parrish.　The Great Plains.
Steedman.　Bucking the Sagebrush.
Bronson.　Reminiscences of a Ranchman.

CHAPTER IX

THE RAILROADS

1. THE PRELIMINARY SURVEYS

Now that we have come to the last chapter of the book, we have also arrived at the last stage in the development of the Great West. That vast expanse of territory lying between the Missouri and the Pacific now becomes a real part of our United States when the iron trail of the locomotive connects the two waters. The millions of people who are on the plains, in the valleys, and on the mountains would not be there had not the railroads pushed into and made a New West.

The buffaloes were the original engineers, as they followed the lay of the land and the run of the water. These buffalo paths became Indian trails, which always pointed out the easiest way across the mountain barriers. The white man followed in these footpaths. The iron trail finished the road.

Lewis and Clark, Pike, Smith, Walker, and Bonneville made valuable explorations. Ashley, Sublette, Bridger, Baker, Carson, and Fitzpatrick explored every stream in the broad domain, hunting for the precious beaver. The Santa Fé, Gila, Spanish, Oregon, and California trails gave well beaten paths through the new land. Lee, Whitman, De

Smet, and Young brought into this wild land some measure of gentleness, and sent from it glowing reports of its richness. Fremont gave to the world reliable maps and reports, making general the geographic knowledge gained in his long

On the Denver and Rio Grande
CAÑON OF THE GRAND RIVER, COLORADO

and difficult expeditions, always giving due credit to that unrivaled transcontinental traveler, guide, and scout, Kit Carson. The gold-seekers from every quarter of the globe dug from the hillsides and washed from the streams that bright metal that caused a stampede westward. The Pony Express flashed across the country facing dangers and storms. The soldier, standing between civilization and savagery, protected the lonely settler and gave his life to

On the Denver and Rio Grande Railroad
ROYAL GORGE, GRAND CAÑON OF THE ARKANSAS, COLORADO
SOME DIFFICULTIES IN MODERN RAILROAD ENGINEERING

redeem the fertile and rich lands of the wilderness. The cowboy came, with his new industry, to feed the millions who had journeyed from afar to build their simple homes in the land of newer freedom and wonderful opportunities.

These were the movements in a western development that now awaited the final act, the building of the railroads. The time had now come when it was necessary to get California and Oregon nearer to the Missouri. The long journey by water around Cape Horn, or by wagon over the trails, was so dangerous and hazardous, and at the same time so long and tedious, that England was actually nearer to the civilization of the States than was the Pacific coast. The Civil War and the finding of gold in the West hastened western development by many years. When the East needed the loyal support of the West during the trying days of the civil strife, and the wealth from the mines was eagerly sought to refill a depleted national purse; when our statesmen began to fear that those Americans so far away, so isolated from the national center, so rich and self-sufficing, might do as the South had done and set up an independent government, no sacrifice seemed too great, no labor too arduous that promised to bind that isolated region to the rest of the Union. We must remember this when we contemplate the generous aid that our country gave to the first builders of transcontinental railroads.

Back as far as the time of Senator Benton, the father-in-law of Fremont, Congress debated the advisability of the government building a road to the West. Benton was always pointing toward the West and crying, "There is the East: there is the road to India." As early as 1835 a proposal was submitted to Congress for building a railroad from the Mississippi to the Columbia along the Oregon Trail; and again, in 1845, Asa Whitney, ever pointing to the Oriental trade, almost succeeded in getting Congress to subsidize

THE BALANCE ROCK, GARDEN OF THE GODS, COLORADO. ONE
METHOD OF TRANSPORTATION

his undertaking to build such a road. In 1850, Stansbury and Gunnison made surveys and explorations in Utah and Colorado; and during the years 1852 to 1854 the government sent out many surveying parties to search for feasible routes across the mountains. In all cases the "paths of least resistance" were adopted. These were the trails made by the Indian, the explorer, the trapper, and the overland home-seeker, no matter which way his journey led, to the north or south, or to the center of the Great West. The old trappers guided the man with the levels and chains to the most accessible passes in the mountains and the level paths of the plains.

Bridger's ears had become well trained by this time, for he could drop a pebble down a deep cañon, and, by counting, could determine how many feet it had fallen when it struck the water. If his ear was well trained, not less so was his eye, for the accuracy of his estimates of distances and elevations was amazing, even to the men who made a practice of this kind of work. The following story is told of his wonderful ability in this line.

"Which of those passes is the lower?" an engineer once asked of him.

"Yon," said the scout, pointing to the south pass.

"I should say they were about the same height."

"Put your clocks on 'em," said Jim, "an' if yon gap ain't a thousand or two feet the lowest ye kin have 'em both."

After a test was made, the south pass proved to be fifteen hundred feet lower than the other one. Bridger would have made a remarkable topographic engineer.

In a government report published between 1855–60, called "Report of Explorations and Surveys to Ascertain the Most Practicable and Economical Route for a Railroad from the Missouri River to the Pacific Ocean," we find five preliminary surveys that resulted as follows: (1) A survey along the 47th and 49th parallels, where now run the Northern Pacific and the Great Northern on lines nearly corresponding to the original survey. (2) A survey between the 41st and 42nd parallels, which does not widely differ from the route of the Union and Central Pacific. (3) A route between the 38th and 39th parallels, which to-day is occupied by the Denver and Rio Grande and the Colorado Midland, parallel roads, and the Western Pacific. (4) A survey near the 35th parallel, covered now by the Santa Fé Railroad. (5) Along the 32nd parallel another survey, now occupied by the Southern Pacific, the extension of the road being from the mouth of the Gila to San Francisco. Take out your geography or map and study the lines of these surveys, and you will find that they correspond wonderfully to the early trails, the path of Smith and Walker, the explorations of Fremont, the caravan roads, and the footpaths of Pattie and Carson.

Before the coming of railroads and fast mail trains, the citizen of Salt Lake, Virginia City, Boise, or Helena received his eastern mail a month, half a year, sometimes a year after it was written or printed. To-day we complain if the morning's mail is a half-hour late. Now the great morning dailies of the city of Chicago are rushed on an overland train that comes puffing and throbbing into the land of the Far West within forty-eight hours from the time the papers were taken from the press. In other words, the people of the West are days, weeks, and even months nearer the eastern centers than they were half a century ago.

The necessity of a transcontinental road became imperative. The settlers demanded adequate protection from the sullen natives, who were now fighting every step that the white man was making toward their old hunting-grounds. Soldiers, ammunition, and provision had to be transported to the West; the settler, with his few belongings, must have better means of travel than that afforded by the oxen, mules, or horses; added to these was the fact that the commerce from the river to the coast and from the Orient to the "Father of Waters" must have quicker and safer transportation.

But the telegraph preceded the railroad. In 1860 Edward Creighton commenced a preliminary survey for a telegraph line from the Missouri to the Pacific. With the usual dangers encountered while going over the trail in the middle of the winter, dangers from highwaymen and Indians, storm and cold, Creighton made the journey in an overland stage coach from Omaha to Salt Lake. From this young city he continued on horseback across the alkali desert and Sierra Nevada to Sacramento, where he interested the California Telegraph Company in his enterprise. Returning to Omaha in the spring, Creighton was ready to commence his undertaking. In the meantime the government had granted a subsidy, a prize of forty thousand dollars a year to the company that should first establish a telegraph line across our new West.

Now Creighton and the California company rushed into the work, each hoping to capture this handsome bonus. Creighton commenced to dig his post holes and string his wires toward the West. The California people worked toward the East. Salt Lake City was the goal, that is, the first company to get its line into that place would get the subsidy. West from Omaha to Salt Lake was a distance of eleven hundred miles, east from Placerville to that point was only four hundred fifty miles. The difficulties to be encountered on the road were about equal. Creighton won, completing his line on the 17th of October, 1861, the California company digging its last hole one week later. When the first message was flashed over the wires on October 24th, another step was made in the process of civilizing the West.

The Indians soon learned that those humming wires carried messages that called for help, and that the soldiers would quickly respond to the call. In their determination to keep the white man back from their territory they tore down the wires by throwing their lariats over the line and then putting their ponies on a run. When this did not do effective work, they took their tomahawks and cut down the poles. The destruction of the telegraph lines was a chief reason for the defeat of our forces along the Bozeman Road. The Indians, just before they made a raid, would destroy the wires so that there could be no communication with the outside world, and the outside world, thinking that if help were needed a call would be made, assumed that all was well.

Even the buffalo seemed to conspire against this last step of civilization. The telegraph poles were rubbed smooth, and sometimes thrown down, by the monarchs of the plains scratching their sides against them. To prevent this destruction, sharp spikes were driven into the poles, but this only added to the delight of the buffalo as the spikes went deeper into the shaggy hair and tough hide.

2. THE UNION-CENTRAL PACIFIC

In order to make possible the building of the first transcontinental railroad, our government appropriated vast sums of money and gave extensive tracts of land to the Union Pacific and Central Pacific railways, the combination of the two roads making a continuous line of transportation from the Missouri to the Pacific coast. In money and land the Union Pacific received $450,000,000, the Central Pacific $380,000,000. Yet with this stupendous gift it was difficult, almost impossible, to get men of means interested enough in the enterprise to risk their money in the venture. Oakes Ames and Sidney Dillon, whose names can never be separated

from that of the Union Pacific, searched the money markets
and pleaded with millionaires for capital to build the great
road. President Lincoln, too, worked night and day trying
to get capitalists to furnish money for the construction of

Union Pacific Railway
GRADING FOR A RAILROAD

the road, which he considered of vast importance. It was
a heartbreaking task. The man of business did not believe
the scheme was practicable. He had no money to invest
in two thousand miles of railway, declaring that the Indians
would tear up the track and make bonfires of the flag stations
along the line. In all the 650 miles between Reno, Nevada,
and Corinne, Utah, there was only one white man living.
How could a railroad pay across such deserts? All honor to
Ames and Dillon for their perseverance. Had it not been

for those two men and the generous aid of Congress this splendid undertaking must have failed.

The first sod of the new road was turned at Omaha on the 5th of November, 1865, and with that act one of the greatest engineering feat of all times was commenced. At the very outstart the company encountered enormous expense. All the machinery, men, cars and material had to be brought up from St. Louis by boats. Wages were high, for each man realized that he was to encounter known and unknown dangers. Then there was no fuel along the right of way. The deserts were treeless, and as a result ties had to be hauled a great distance. Often there was no stone or rock for ballast. In fact, as Warman says, "They found absolutely nothing; only the great right of way and the west wind sighing over the dry, wide waste of a waveless sea."

"For three long winters engineers living in tents and dugouts watched every summit, slope, and valley along the entire fifteen hundred miles of road, to learn from the currents where the snow would drift deep and where the ground would be blown bare. In summer they watched the washouts that came when the hills were denuded by what, in the West, they called cloud-bursts. These were the only experts competent to say whether a draw should be bridged or filled, and only after years of residence in the hills."[1]

The story of the Indian depredations committed along this road while it was being built would fill many books. Loss of life and property was a daily occurrence. These trail-makers were as brave and fearless as those who had gone over the path years before. At some points along the line they never knew when night came if they or some one else would be wearing their scalps in the morning. A large number of the men employed in the construction of the road

[1] Warman. The Story of the Railroad. Copyright, 1898. D. Appleton & Co.

DALE CREEK BRIDGE (Wyoming)

The bridge (25 feet high and 500 feet long was completed in thirty days.

had been soldiers in the South and easily adapted themselves to camp life and the dangers of the plains. These unemployed men who had experienced battles in the ranks of war found, civilization too tame for them, and longed for some excitement. ᾽ The freedom of the West offered alluring attractions, and the railroad construction benefited by this restlessness. Thus, at a moment's warning, a thousand men could be put into the field thoroughly trained for a battle, able to subdue three times their number of untrained and undisciplined red men. In the ranks of these railroad employees might be found those soldiers of other battles who had ranked from generals to privates.

Buffalo Bill obtained his name from the occupation he followed at this time. When the Union Pacific Railroad was constructing its line toward the West, in 1867, Cody was hired to furnish fresh buffalo meat to the construction men. On account of the hostility of the Indians it was an extremely dangerous undertaking, but Cody undertook this task. When this hunter would come into camp of an evening with his wagon filled with fresh meat, the men would exclaim, "Here comes old Bill with more buffalo," until the two names Bill and buffalo, became inseparable.

During one year Buffalo Bill killed for the railroad four thousand two hundred eighty buffalo with the help of his pony, "Old Brigham," and his breach-loader, "Lucretia Borgia."

It is to be observed that when the North Platte was reached there was a "parting of the ways," for at this station the Union Pacific left the Oregon Trail, going south and then directly west in place of following the river to Fort Laramie. There are those to this day who believe that the better route would have been along the old highway and through South Pass, but to the south the Indians were less troublesome and the roadbed was more easily constructed.

With the commencement of the Union Pacific at Council Bluffs, the Central Pacific began its construction work at Sacramento, California, at the western limit of the wilderness. This part of the road was pushed forward with feverish haste toward the East and met the west-bound road

From an old Photograph
THE MEETING OF THE UNION AND CENTRAL PACIFIC RAILWAYS
AT PROMONTORY POINT, UTAH, MAY 10, 1869

in Utah. There, at Promontory Point, about thirty miles west of Ogden, on the tenth day of May, 1869, the final connection of the two lines was made with great ceremony amidst the joyful plaudits of more people than one could expect that bare region to furnish. With a silver hammer four spikes were driven at the union of the railroads, two silver and two gold spikes, coming from the mines of Montana, Nevada, California, and Idaho. With this ceremony came speeches and prophecies from the great

Union Pacific Railway

AN OVERLAND LIMITED IN FULL SWING

leaders who had pushed this gigantic work to its completion.

When George Francis Train made a rousing speech in Omaha at the time of "the breaking of the sod" he was laughed at for saying that the road to the West would be

Union Pacific Railway
THE IRON TRAILS, SHOWING THE BALLASTED ROADBED, DOUBLE TRACK AND ELECTRIC BLOCK SIGNALS

completed in five years. He was called a dreamer. The actual time was three years, six months, and ten days. Train missed his guess by a year and a half.

The Union Pacific had built eleven hundred eighty-six miles from the Missouri, the Central Pacific six hundred thirty-eight miles east of Sacramento.

No wonder that there was rejoicing on this historic day, for it was the beginning of a new era, and an infinitely better era for both East and West. The West saw in it the beginning of the end of hardship and privation; the East saw a new and wonderfully rich land waiting to be exploited.

LUCIN CUT-OFF. A RAILROAD ACROSS GREAT SALT LAKE

The Pacific railroad was completed, and the day of the trapper, the explorer, the Pony Express, and the emigrant wagon over the old trails was no more, for the iron trail had come into possession of its own.

3. THE SOUTHERN PACIFIC AND THE ATCHISON, TOPEKA AND SANTA FÉ RAILROADS

When the Central Pacific was constructed the engineers encountered many serious difficulties from the heavy fall of snow and the drifts that blockaded the trains. Many miles of wooden snowsheds had to be built, particularly in the Sierras. To the expense of the original building of these sheds was added the great cost of replacing those that were burned by sparks from the locomotives, for the wooden constructions were very dry and easily caught fire. The owners of the Central Pacific, Huntington, Stanford, and Crocker, wishing to avoid the snowdrifts in the north and the constant repairs on the sheds, sought a southern route and built the Southern Pacific Railroad, which started from San Francisco and ended in El Paso, Texas.

The Gadsden Purchase was made by our government in 1854 for the sum of ten million dollars, in order to have a territory through which to construct a railroad. Jefferson Davis, in 1853, then Secretary of War, had five surveys made for transcontinental roads. He naturally favored the most southern one on the 32nd parallel. But the best route was that followed by Brigadier-General Cooke and the Mormon Battalion in taking through General Kearny's wagons in 1847. Therefore the Gadsden Purchase was made, whereby our government obtained parts of the Mexican states of Sonora and Chihuahua, and had an elevated table-land on which to run a southern railroad, avoiding the

Southern Pacific Railway

SNOWSHEDS (California)

Rocky Mountains and the Sierra Nevada Mountains with their deep snows and steep slopes.

The Gadsden Purchase was denounced, and the expenditure of such an enormous amount of money for only forty-five thousand acres of cactus desert, unfit for cultivation, was declared a waste of money. But this land was valuable from a strategic standpoint as a passageway to the coast, and its purchase forever settled the boundary question between the United States and Mexico.

The government not only gave the Union-Central Pacific vast tracts of land, but the Southern Pacific came in for its share with twenty-four million acres and the Santa Fé system received seventeen million. Even with this generous help it was a huge undertaking and only men of large views dared to risk their money in either road.

The line of the Atchison, Topeka and Santa Fé followed the old trail to a remarkable degree. The Santa Fé caravan made a roadbed that was often used for the iron trail.

A few miles north of Topeka is the little town of Wakarusa, at which point the railroad follows the route of the caravan trail. Even where the routes do not coincide they run so close together that nearly every stream, hill, and wooded dell recalls some story of adventure that occurred in those days when the railroad was regarded as an impossibility and the region beyond the Missouri was a veritable desert.

At Pawnee Fork the railroad crosses the river at exactly the place that the old trail did. Here were fought some of the bloodiest battles of the plains between the hostile tribes. Here the Indians also fought freight wagon, coach, and every kind of outfit that passed over the trail, for this was the natives' greatest resort for robbery and murder.

The railroad passes by the ruins of Fort Bent and the old camping-grounds of the Arapahoes and the Cheyennes. It crosses Raton Pass precisely as the old trail did, and just before reaching the summit it passes the tavern of Uncle Dick Wooton, who built the toll road over the pass. Uncle Dick's old buildings were to be seen at the right of the track by all going westward on Santa Fé trains until two years ago; but they have been torn down to make way for a handsome mountain lodge, for J. Pierpont Morgan has bought thirty-two thousand acres in that region and has turned it into a coal mining and hunting ground.

In February, 1880, the railroad reached the sleepy old city of Santa Fé and the day of the ox team, the caravan, and the stage coach was over. The Santa Fé Trail had passed into history.

On the Atchison, Topeka and Santa Fé Railway

MODERN FREIGHTING TO SANTA FE

The largest locomotive in the world, called a mallet articulated compound engine. Weighs about 850,000 pounds.

About the time that the Atchison, Topeka and Santa Fé Railroad reached the city of Santa Fé, the Southern Pacific had reached El Paso. Now, it was the intention of the former railroad to push its line to the Pacific coast as soon as possible and it was also the purpose of the other road to construct a line east to the Mississippi or the Gulf of Mexico. By mutual agreement, however, these two roads met at Deming, near El Paso, and used each other's lines to reach from the coast to the Mississippi, and thus was finished the second transcontinental line. Ultimately the Southern Pacific built east as far as New Orleans, and pushed a line northward from San Francisco to Portland that ran parallel with the trail of Jedediah Smith, when he went from Monterey to the home of Dr. McLoughlin at Vancouver. Then the Santa Fé finally constructed its own line to San Francisco, and again the Missouri and the Pacific were connected.

4. THE NORTHERN PACIFIC

The Northern Pacific Railway Company received a charter in 1864 to build a railroad from some point on Lake Superior, in the state of Minnesota or Wisconsin, westward on a line north of the forty-fifth degree of latitude to a point at or near Portland, Oregon. Furthermore, it received a land grant of forty sections of public land per mile of road through the territories, and twenty sections through the states across which its road would be constructed.

The beginning of the construction of the Northern Pacific was made near Duluth. The equipment resembled the materials for a military campaign more than for a peaceful survey. Almost every mile Indians were encountered, who contested every foot of progress. These first preliminary surveys were made in the early fifties. The same problem of obtaining money that had confronted the other roads had to be met by this one, but construction work began in the summer of 1870. Money was not to be had, however, and the road went into the hands of a receiver in 1875. Then came Henry Villard, the real genius of the Northern Pacific. He had extensive transportation interests on the Columbia, thoroughly appreciated the wonderful possibilities of Washington and Oregon, the home of the farmer and the settler with room for hundreds of thousands more, and he took hold of the bankrupt road and pushed it on to the coast.

It took thirty years to build the Northern Pacific, and like the Union, Central and Southern Pacific and the Santa Fé, it was obliged to receive government aid for its construction. No private fortune at that time could be expected to undertake the gigantic task of putting a railroad across deserts, over mountains, through rocks and across tremendous rivers just to connect the Pacific with the Atlantic. The "last spike" ceremony enacted September 8, 1883, in western Montana, was second in interest only to that of the Union-Central Pacific. When Villard hammered down the last spike amidst the noise of loud cheering, booming of artillery and strains of martial music, he was surrounded by prominent men not only of America but also of Europe.

This Northern Pacific sweeps over the rich prairies of Minnesota, through the most fertile farm lands of North Dakota, and stretches onward to the Yellowstone, up this branch of the Missouri, through the towns of Billings, Livingston, and Bozeman, above which one line branches off to Helena while another runs through Butte, both branches reuniting farther west to run through Missoula, then over the Bitter Root Mountains, the divide between Montana and Idaho, across the handle of Idaho to Spokane, Washington, then south to the historic Columbia at Pasco, and along the great river to the railroad's terminal at Portland, whence extensions rush to Seattle and Tacoma. A wonderful pioneer road through a more wonderful Northwest, directly across five states that furnish the world's market with grain, gold, silver, copper, cattle, sheep, fruit, timber, and fuel in tremendous quantities.

The Northern Pacific Railway is intimately related to the Lewis and Clark trails. "From Bismarck and Mandan on

From Northern Pacific Railway
IRRIGATION CANAL IN BITTER ROOT VALLEY

the Missouri River in North Dakota, it parallels the explorers' line of travel along the Missouri, Yellowstone, Gallatin, and Jefferson rivers to Helena and Whitehall, and, on a part of the Hellgate River to Missoula, Montana, its main lines again connect with it on the Columbia River in eastern Washington and also in Oregon. Through its branch lines it meets or parallels the route in the Jefferson, Bitter Root, Clearwater, and Snake rivers in Montana, Idaho, Washington, and Oregon."[1]

[1] Wheeler. The Trail of Lewis and Clark. G. P. Putnam's Sons.

Thus the Pacific is reached for the third time, and the trail of the explorer has been crossed, united and recrossed by the iron trail, just as it was by the other two transcontinental lines.

5. THE GREAT NORTHERN

It sometimes has been claimed that the Great Northern Railroad is the only transcontinental line that was built without public aid. While this is almost true, it is not quite an accurate statement. When James J. Hill, the famous railroad man of St. Paul, in 1879, organized the St. Paul, Minneapolis and Manitoba Railroad it received 3,675,000 acres of free land from the United States. With the exception of this subsidy, that most northern of our railroads was built with private capital. The Great Northern Railroad was an outgrowth of the St. Paul, Minneapolis, and Manitoba, and was completed in 1893, running from St. Paul to Seattle.

Hill may well be called one of the great captains of the world, for he pushed his line to the coast, depending entirely upon the growth of the country through which his road ran and the trade he expected to have with Asia and the Pacific Islands to repay him. For a time he had in operation a fine line of steamers, some of them the largest in the world, running from his terminus on Puget Sound to China, Japan, Honolulu, the Philippines, and even to Siberia. These vessels carried over grain, flour, and machinery, and brought back fruits, silks, tea, spices, and fancy wares.

A century ago, if you remember, John Jacob Astor had a cherished dream of doing this very thing,— making a transcontinental route and at the end of it a line of ships to trade with the Orient. Hill realized this dream with his Great Northern.

This road made a new record for speed in construction

from Fargo, North Dakota, to Helena, Montana. Material for miles of equipment was constantly rolled forward to the builders. "The supply train was unloaded in a drilled confusion of mad haste near the end of the track. Ties and rails were seized as soon as they touched the ground and were hurled to the front by galloping horses; and the system

BRIDGE OVER THE MISSISSIPPI. HASTINGS, MINNESOTA

was so elaborately studied that five hundred seventy blows an hour was an exact standard of performance for each spiker."

Lewis and Clark made some of the preliminary surveys also for this road. The trail of the explorers is virtually paralleled from Minot to Helena. The Great Northern passed up the Maria's River to the exact spot where Lewis had his first and only combat with the Indians. The southern branch of the road, going south past Fort Benton and the Great Falls, follows the old trail where the iron boat was cached in 1805 and put into use again in 1806 when Lewis returned from the coast. The iron trail follows the portage trail of the expedition, passing too far away to give a view of the Great Falls, but permitting passengers to view from a palace car window the Rainbow and Black Eagle falls above, which the explorers saw after such a toilsome and sore-footed portage over a century ago.

When the Great Northern was to be put across the continent men laughed at Hill's ambition. The idea of building a road north of the Northern Pacific, which they thought already so far north that the country through which it ran could not grow wheat, seemed absurd, and Hill's new

On Denver and Rio Grande Railroad
CURECANTI NEEDLE, BLACK CAÑON OF THE GUNNISON. COLORADO

venture was called "Hill's Folly." But Hill had faith in himself and in the country through which he was constructing his road. He knew that land and what countless thousands it was capable of supporting. He looked into the future and saw his line running through a populous region of fine farms, the future wheat belt of the world, where thriving villages and busy cities would have transportation needs to try the best equipped road. And he set himself to the task of building up this region in the shortest possible time. He sent his agents all over northern Europe, telling of the fertile valleys, the extensive irrigation projects, the unrivaled advantages of farm life in Montana, Idaho, and Washington, and the

Scientific American
A BIPLANE RACING WITH A TRAIN

ease of acquiring a comfortable home. His seed fell on good soil and thousands of honest, hard-working Germans, Swedes, Norwegians, Danes, Scotchmen, Englishmen, Irishmen, yes, and native Americans, are pouring into that country and building up in our Northwest commonwealths second to none in the Union. All along the line of this great railroad towns now exist where only a few years ago no one lived at all; small settlements have grown into prosperous cities; and lands but lately barren have become vast wheat belts furnishing bread to all countries of the world.

While these great trunk-lines crossed the continent from the river to the coast other lines were run to the South and up from the South, and spurs and side lines were added to this network until almost every part even of the Great West has become fairly easy to reach.

Witness the growth of the West. In a few years the barren wilderness has changed into a prosperous land of homes and schools; the cactus of the prairies has given place to fields of grain; the desert plains have grown fertile through the use of the irrigating ditch; the grazing lands of the buffalo have been turned by traction plows; and all things have changed except the sunshine, the singing winds and the everlasting hills.

While we are enjoying the luxuries of this new era of the Great West let us not forget to honor those who endured hardships and privations, encountered dangers and peril; yes, even gave up their lives to make these things possible. Let us ever remember Lewis and Clark and Pike, who blazed the way; the grizzled old beaver hunters, who explored every nook and corner and mapped out routes; the trader and the caravan merchant, who with their wagons wore down the trails and made them possible for colonists; the scientific explorers, who spread authentic accounts far and wide among the people of the East; the missionaries, who braved death to spread the Gospel of Christ; the gold-seekers, who came in crowds to the mountain sides and started town life; the soldiers, who fought and died to protect the home-seeker, his wife and babies; the picturesque cowboy, who once ruled the plains; and the railroads, which came to bind inseparably the East and the West and to fill the plains and mountain lands with fruitful farms and thriving towns, and, lastly, let us not forget the Indian, the original native whose tragedy underlies the white men's triumph.

It is all a story that has never had its equal in the world's history. The Great American Desert is no more. "The West has changed. The old days are gone. The house dog sits on the hill where yesterday the coyote sang. The fences are short and small, and within them grow green things instead of gray. There are many smokes rising over the prairie, but they are wide and black, instead of thin and blue." [1]

REFERENCES

Warman. The Story of the Railroad.
Spearman. The Strategy of the Great Railroads.
Wheeler. The Trail of Lewis and Clark.
Dellenbaugh. Breaking the Wilderness.
Laut. Story of the Trapper.
Inman. The Old Santa Fé Trail.
Paxson. The Last American Frontier.

[1] Hough. The Story of the Cowboy. (Copyright, 1897. D. Appleton & Co.)

PRONOUNCING VOCABULARY

Arikara, a rĭk a ra
Arrastre, ä räs'tre
Assinnibonne, a sin'i-boin

Baptiste, bă teest'
Bonneville, bon'vil

Cache, kash
Cameahwait, kăm'e ah wait'
Cpaparejosare, shă pa rê ho sä're
Chaps, shaps
Charbonneau, Toussaint,
   shär bon no, too sang
Chihuahua, chē-wa'wa
Cheyenne, shi ĕn'
Chopunnish, cho-pun'ish
Chouteau, Pierre, shoo to, pē ar
Clatsop, klat'sop
Coboway, co bo'way
Comcomly, kom kom'ly
Cibola, sē'bo-lä
Creighton, cray ton
Cimarron, se-ma-ron'

Dalles, dălz
De Cardenas, da kar'-da-nas
De Vaca, Cabeza,
   da vä ka, ka-va'thä

Forsyth, for-sīth

Gila, he'-lä
Gros Ventre, gro vŏnt'
Guadalupe Hidalgo,
   gwä-thä-lo'pä-ē-däl gō
Geronimo, hē ron'i mō

Hidatsa, hē-dä'-tse
Junipero Serra, hu ni per'o ser ra
Kalispel, kăl is pĕl'
Lajeunesse, Basil,
   la-zhe-nes', bā'zil
La Ramie, Jacques,
   la ra'me, zhäk
Lisa, le sah

Manuel, man'wāle
Marbois, Barbe, mar-bwa', barbe'
Mer de l'Ouest, mār del west
Michaux, mē-sho'
Monterey, mon te ray'
Minnetarees, min-e-tar ees'

Natchitoches, nach i toch'ez
Nez Perce, nā'par'sä

Ogalalla, o-gal läl'lä
Orofino, oro feéno

Padre, pä'dre
Pend d'Oreilles, pon doray
Piegan, pe'gan
Pierre's Hole, pē ars'
Preuss, prois
Provost, Etienne, pro vo', ā-tĕn

Quivira, kē-vē'-rä

Rio Grande, rio grand
Rendezvous, ren de voo
Sacajawea, sak-a-jâw'-é a
San Jose, san ho zä'

255

San Juan, san hwän'
Shoshone, shō-shō'ne
Sublette, sub let
Sioux, soo

Teton, te'ton
Ten Eyck, ten ike'
Tiolizhilzay, te ol i zil' zā
Tlamath, klä'mat
Tonquin, ton keen'

Villard, vil lard'
Verendrye, vär en'dree

Wasatch, wa'-sach
Willamette, wil am'et
Wyeth, y'-eth

Yosemite, yō sem'i te

Zuni, zoon yee'

BIBLIOGRAPHY

*ADAMS. The Log of a Cowboy.
*ALTSHELER. The Last of the Chiefs.
*ALTSHELER. The Horsemen of the Plains.
BANCROFT'S Histories. Separate volumes on the various States West of the Mississippi.
BECKWOURTH. Life and Letters.
BELDEN. The White Chief.
*BENNETT. A Volunteer with Pike.
BOWLES. Across the Continent.
*BOURKE. On the Border with Crook.
*BOYLES. The Spirit Trail.
*BRADY. The Conquest of the Southwest.
*BRADY. Northwestern Fights and Fighters.
*BRADY. Indian Fights and Fighters.
BRIGHAM. Geographic Influences in American History.
*BRONSON. Reminiscences of a Ranchman.
BROOKS. First Across the Continent.
*BROWN. The Glory Seekers.
BRUCE. The Romance of American Expansion.
*BURNETT. Recollections of an Old Pioneer.
CARRINGTON. Army Life on the Plains.
*CARTER. When Railroads were New.
CHITTENDEN. The American Fur Trade.
CHITTENDEN. The Life of Father De Smet.
*CODY. The Tales of the Plains.
COOKE. The Conquest of New Mexico and California.
COUES. The Expedition of Zebulon Pike.
COUTANT. History of Wyoming.
CUSTER. Life on the Plains.
*CUSTER. Boots and Saddles.
DELLENBAUGH. Breaking the Wilderness.
*DYE. McLoughlin and Old Oregon.

* To be used for reading rather than for reference.

*Dye. McDonald of Oregon.
*Dye. The Conquest.
Eels. Marcus Whitman.
Fairbanks. The Western United States.
Finerty. Warpath and Bivouac.
*Forsyth. The Story of the Soldier.
Fremont. Story of my Life.
Gregg. Commerce of the Prairies.
*Grey. The Last of the Plainsmen.
Grinnell. The Story of the Indian.
*Grinnell. Trails of the Pathfinders.
Hailey. History of Idaho.
*Hansen. The Conquest of the Missouri.
Hebard. History and Government of Wyoming.
Herman. The Louisiana Purchase.
*Hitchcock. The Louisiana Purchase.
*Hosmer. The History of the Louisiana Purchase.
*Hosmer. The Expedition of Lewis and Clark.
*Hough. The Way to the West.
*Hough. The Story of the Cowboy.
*Irving. Astoria.
Irving. The Adventures of Captain Bonneville.
Inman. The Great Salt Lake Trail.
Inman. The Old Santa Fé Trail.
*Jackson (Helen Hunt). Ramona.
*Johnson. Pioneer Spaniards in America.
*Judson. Montana.
Knower. The Day of the Forty-Niner.
Langford. Vigilante Days and Ways.
Larpenteur. Forty Years a Fur Trader.
Laut. The Story of the Trapper.
*Laut. The Pathfinders of the West.
Lewis and Clark. Journals.
*Lummis. The New Mexican David.
Lummis. The Spanish Pioneers.
Lummis. Pioneer Transportation in America.
Lyman. The Columbia River.
*McBeth. The Nez Perce Indians since Lewis and Clark.
*McLaughlin. My Friend the Indian.
*McMurray. Pioneers of the Rocky Mountains.

       * To be used for reading rather than for reference.

*McNEIL. With Kit Carson in the Rockies.

MAJORS. Seventy Years on the Frontier.

MARCY. Army Life on the Border.

MARSHALL. History vs. the Whitman Saved Oregon Story.

MEANY. History of the State of Washington.

MILES. Personal Recollections.

MONTANA. Contributions to the Historical Society.

PAINE. The Greater America.

PARKMAN. A Half-Century of Conflict.

*PARKMAN. The Oregon Trail.

*PARRISH. The Great Plains.

*PARRISH. Keith of the Border.

*PAXSON. The Last American Frontier.

PRINCE. Historical Sketches of New Mexico.

RICHARDSON. Beyond the Mississippi.

*ROOSEVELT. Ranch Life and the Hunting Trail.

ROOSEVELT. The Wilderness Hunter.

ROOSEVELT. Winning of the West.

*ROOSEVELT. Stories of the Great West.

*ROOT and CONNELLEY. The Overland Stage to California.

ROYCE. History of California.

SCHAFER. History of the Pacific Northwest.

SEMPLE. American History and Its Geographic Conditions.

SPEARMAN. The Strategy of the Great Railroads.

STANSBURY. Report of Great Salt Lake.

*STEEDMAN. Bucking the Sagebrush.

*TALBOT. My People of the Plains.

THWAITES. Early Western Travels, a large number of Journals edited
by Dr. Thwaites.

*THWAITES. Rocky Mountain Explorations.

*TWAIN. Roughing It.

TWITCHELL. The Military Occupation of New Mexico, 1846–1851.

*VISCHER. The Pony Express.

WARMAN. The Story of the Railroad.

*WETMORE. The Last of the Great Scouts (Buffalo Bill).

*WHEELER. The Trail of Lewis and Clark.

WHITNEY. History of Utah.

WINSHIP. The Journey of Coronado.

*WISTER. The Virginian.

*WOOD. The Lives of Famous Chiefs.

* To be used for reading rather than for reference.

INDEX

Printed in the United States
40315LVS00004B/224